Fundamentals of Environmental Sampling

Keith Bodger

Government Institutes
An imprint of
The Scarecrow Press, Inc.
Lanham, Maryland • Toronto • Oxford

Published in the United States of America
by Government Institutes, an imprint of The Scarecrow Press, Inc.
A wholly owned subsidiary of
The Rowman & Littlefield Publishing Group, Inc.
4501 Forbes Boulevard, Suite 200
Lanham, Maryland 20706
http://www.govinstpress.com/

Estover Road
Plymouth PL6 7PY
United Kingdom

British Library Cataloguing in Publication Information Available

Library of Congress Cataloging-in-Publication Data

Bodger, Keith.
Fundamentals of environmental sampling / Keith Bodger.
p. cm.
Includes index.
ISBN 13: 978-0-86587-957-7 (pbk. : alk. paper)
ISBN 10: 0-86587-957-5 (pbk. : alk. paper)
1. Environmental sampling. I. Title.

GE45.S25 B64 2003
628'.0287—dc21 2003045224

The paper used in this publication meets the minimum requirements of American National Standard for Information Sciences—Permanence of Paper for Printed Library Materials, ANSI/NISO Z39.48-1992. Manufactured in the United States of America.

For Mel, my lovely wife.
Thank you for your support throughout this endeavor
and the myriad steps leading to it. I love you.

Contents

List of Figures and Tables

Preface

This book is intended to be a supplement to the many texts and standard practices that tell you how to sample. There are numerous ASTM standards and guidelines that address sampling methods. In addition, the EPA addresses numerous sampling methods and technologies, which may be found at their web site. This book should not replace them, but rather complement them. It does not intend to teach you how to sample, but rather why we sample and how to make it easier. Throughout this book, I will occasionally share with you some of the mistakes I have made in my sampling career, in the hope that you may avoid making the same mistakes I did. As someone once said, "Learn from other people's mistakes; there isn't enough time to make them all yourself."

Chapter 1, Fundamentals of Geology, explains how to describe the samples you collect when you are in the field. It provides an overview of several lab procedures on which soil descriptions are based. With this practical understanding, you should realize the importance of accurate sample descriptions. This chapter also explains the importance of completing a borehole log and how engineers and geologists use the information.

Chapter 2, Fundamentals of Chemistry, presents the chemistry of sampling in simple terms, so that you understand why you are collecting samples and how the results affect the overall goal of the project. The various sample analytes are discussed, as are how these analytes are decided and where to find the regulations that guide these analytes. The purpose behind collecting each parameter is presented to give you more of a purpose for going in the field.

Chapters 3 through 6 discuss the methods used in the field. I will explain the advantages and disadvantages of the methods, where to find guidelines for the methods, and tips on how to prevent mistakes when using these methods. I will also discuss how each method affects sample results. These chapters are practical, informational, and easy to read. Having spent 10 years in the field, I have a good idea of what needs to be discussed and how your efforts affect the overall goals of the project.

The goal of environmental sampling is to collect a valid sample that accurately represents conditions at the site. Chapter 7 focuses on how to collect valid data and how the various "extra" samples are used to validate data. I will discus sample preservatives, why they are used, and what they do. I will also discuss the data validation process, which is conducted in the lab and often in the office.

Having gone into the field too many times without the right equipment, I provide a list of the right tools to do the right job in Chapter 8. I explain why the tools are needed, how they are used, and how to store them so that you can find them. I also talk about tools that you carry with you on a daily basis (e.g., hands and feet) and practical ways to use them in the field.

Finally, Chapter 9 shows how to bring it all together in the office, and how data collected in the field are used in the office. I discuss the written report, client needs and points of view, and relationships in the field and in the office.

Throughout the book I provide career tips on how to advance in this field. There are tips that I have followed, as well as tips I have not followed and should have. Using this book should give you some great insight that will help you throughout your environmental career.

About the Author

Keith Bodger is an Environmental Specialist at one of the largest natural gas distribution companies in the country. He has been an environmental sampler for over ten years. He currently maintains compliance of more than 40 underground storage tanks, more than 70 aboveground storage tanks, and manages multiple remediation projects. He has an Associates Degree in Geotechnical Engineering from Sir Sandford Fleming College in Lindsay, ON and a B.S. in Environmental Studies from Northwestern University in Evanston, IL. Mr. Bodger is a Certified Hazardous Materials Manager and lives in Wheaton, IL with his wife and two kids.

Acknowledgement

No sampler is an island; therefore there are many whom I must thank for their help in this book. First, thanks to Rich Trzupek at Huff & Huff, Inc., who wrote Chapter 6, Air Sampling. Rich has the uncanny ability to make the complex air regulations understandable to those who merely use air for breathing, like me. Thanks to Doug Dorgan at Weaver Boos Consultants, Inc., who provided a technical review of the text. While working for Doug, he taught me to write and to understand the importance of accuracy and consistency. Without my having worked for Doug, this book would not have been possible. Thanks to Laurie Franklin at First Environmental Laboratories, who provided information and clarity in the Chemistry Fundamentals and Valid Data chapters. Thanks to my editors, Will Bradbury and Charlene Ikonomou at Government Institutes. Your patience is appreciated, as is this opportunity. Thank you to Wood's Haven on the Rock, a wonderful environment in which to write a book. I want to thank the many people with whom I have sampled. Thanks for the laughter, the songs, the trivia, and the stories. You helped to make the hours seem like minutes. Finally, thanks to all of you who taught me how to sample. Your influence is scattered throughout this book.

1

Geology Fundamentals

It is crucial in environmental work to have a basic understanding of geology. While the industry buzzword is "environmental," one-third of the environment is geology—the other two-thirds being water and air. And even water is dependent upon the geology over or through which it travels. So, geology is important. To explain it, I will first discuss the rocks from which soil comes, and then the characteristics of soil.

Rock

A rock is two or more minerals (e.g., granite is primarily orthoclase feldspar and quartz). A mineral is something with a fixed chemical composition (e.g., quartz is SiO_2 or silicon dioxide). There are three types of rocks: igneous, metamorphic, and sedimentary. Igneous rocks are a result of either volcanic activity (e.g., lava (molten rock) flowing from a volcano) or from magma (molten rock that is in the earth). Examples of igneous rocks include granite, diorite, and basalt. Sedimentary rocks are, as the term suggests, formed from sediments or particles. These particles lithify (harden) together to create a rock. The particles may be physical parts from other rocks such as sandstone (pieces of sand) or they may be a result of a chemical reaction (a precipitate) such as limestone.

Sediments are created primarily through weathering processes including water, ice, heat, and wind. From physical weathering we get sandstone (pieces of sand lithified together). The sand breaks away from its parent rock through water or wind effects, typically. Chemical weathering affects substances that can be held in solution (e.g., salt, calcium). From chemical weathering we get limestone and dolomite (formed by calcium carbonate in solution).

Finally, metamorphic rocks are formed from changes in rock formations caused by heat or pressure from overlying rock and sediment. Sandstone, when subjected to heat and pressure will become quartzite. When subjected to the same conditions, shale will become slate and granite will become gneiss. Extreme pressure will also deform rock resulting in folds and bends. Soil is either a physical or chemical breakdown of the three rocks above.

Soil

When there is a physical breakdown of the rock, we get soil in the form of boulders, cobbles, gravel, sand, and silt. When there is a chemical breakdown, we get clay particles. During environmental sampling, four sizes of soil will be typically encountered: gravel, sand, silt, and clay. We do not typically collect cobbles or boulders for chemical analysis.

Soil Size

Before you can accurately identify soil in the field, you must have an understanding of soil size and how it's determined. Of the four sizes of soil, gravel is the largest and clay is the smallest. Each of these particles has a defined size. For example, gravel measures less than 76.2 mm (3-inches) and greater than 4.74 mm (0.19 inches). Sand measures between 4.74 mm (0.19 inches) and 75 µm (0.003 inches). Silt is between 0.075 mm (75µm) and 0.002 mm (2µm), while clay particles measure less than 2 µm. Particle sizes are measured not by tape measure, but by using sieves for gravel and sand, and a hydrometer for silt and clay (see Figures 1.1 through 1.4).

Figure 1.1. #4 Sieve (4 wires per inch). If the particle falls through this, it's sand. If it doesn't fall through (retained), then it's gravel.

Figure 1.2. 3/8″ Sieve. This is the smallest gravel sieve.

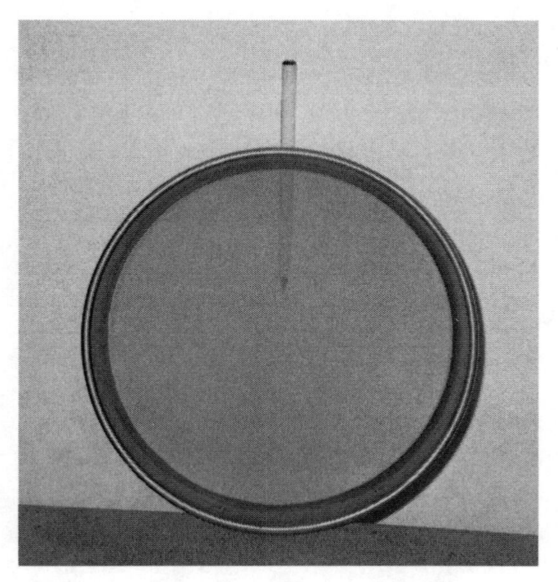

Figure 1.3. #200 Sieve. If a particle falls through this, it's silt or clay. If it's retained on this sieve, then it's a fine sand.

Figure 1.4. A nest of sieves. A soil sample is placed in the top and the soil falls through the various sieves. Soil retained on each sieve is weighed and plotted on a graph.

The standard way of measuring the size of particles is using a nest of sieves (several sieves stacked on top of each other) (see Figure 1.4). Keeping in mind the concept of a kitchen sieve (small particles fall through, large particles stay in the sieve), a nest of sieves is a series of brass rings each with a grid or mesh in its base through which soil will fall, depending on the particle size. The smallest gravel sieve has openings that measure 3/8 of an inch while the largest has openings that measure three inches. There are larger sieves for larger particles. For sand and smaller, the size of the sieve is based upon the number of threads-per-inch in the base of the sieve. This thread-per-inch concept is the same as when you buy sheets for your bed. Sheets typically come in counts of 180, 200, 220, or 250 threads-per-inch. The higher the thread count the higher the quality, durability, and cost. When you eat crackers in bed, if the crumbs are fine enough, they will fall through your sheet as a very fine sand (180 thread-count) or a silty clay size (greater than 200 thread-count). The largest sand sieve is a #4 sieve, meaning it has four threads-per-inch. You might think that the sand particles measure ¼ of an inch, but that does not account for the diameter of the thread. Therefore, the largest sand particle is less than 1/5 of an inch (0.187 inches). The smallest sand sieve is a #200, which has 200 threads-per-inch. The human eye cannot see particles smaller than a #200 sieve.

To measure particle sizes smaller than a #200 sieve, it is necessary to conduct a hydrometer analysis. While the methodology behind a hydrometer analysis (ASTM D-422) is far beyond the scope of this book, the general concept is to take a measurement of soil that passes a #10 sieve, mix it thoroughly with distilled water and an anti-coagulant, and then place it in a tall glass cylinder. A hydrometer bulb (like an over-sized thermometer) is placed in the mixture and you measure the hydrometer as it sinks. As the particles settle to the bottom, the hydrometer sinks lower into the mixture. The silt and clay particles take much longer to settle, so this test actually lasts 24 hours.

The purpose of the sieve analysis is to determine the weight of each soil size in your sample. Once you place the soil in the sieve nest, shake it, then weigh the soil on each sieve. That weight is plotted on a semi-log graph and a line is drawn to "connect the dots." With the fine portion (passing #200), the hydrometer reading and time are computed (usually with a simple lab software program) and those data are plotted on a semi-log graph (see Figure 1.5). The result is a "curve" that represents the soil sample. This paragraph just replaced (albeit insufficiently) about three two-hour soil labs.

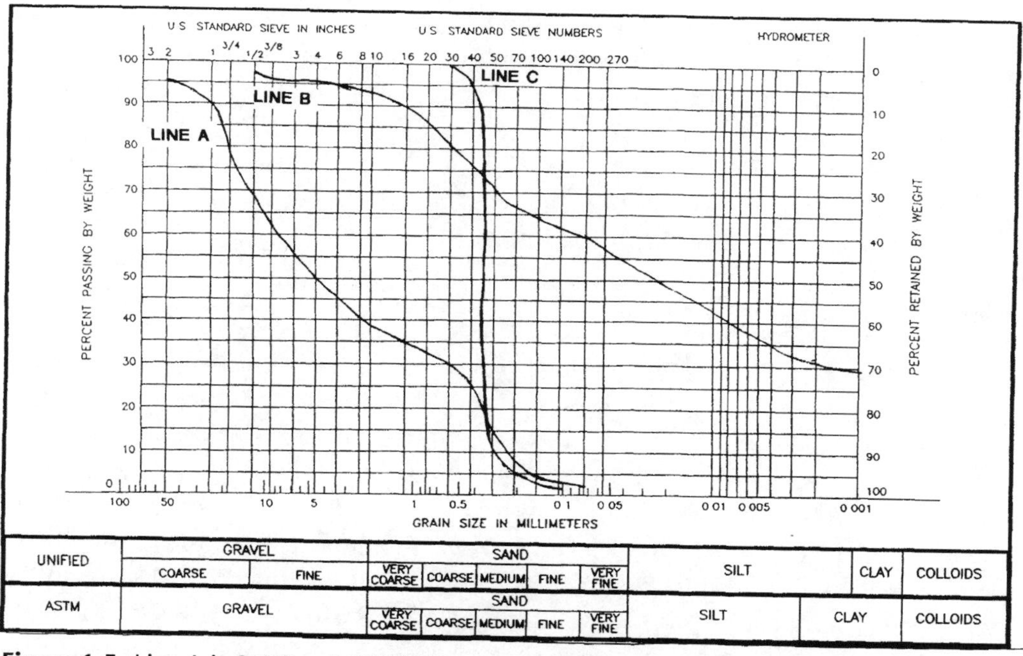

Figure 1.5. Line A is SAND and GRAVEL. Line B is CLAYEY SILT, some sand, trace gravel. Line C is course to medium SAND (perhaps typical of a sand trap at a golf course).

Conducting a sieve analysis can provide several key characteristics that are quite obvious when looking at the curve. For example, if the curve is long and gradual (see Figure 1.5), it is well graded. That means there are soil particles at each sieve size. If it is a short curve and drops down to the #200 sieve quickly, then it is well sorted. This means there are few soil particles on several sieves. Well-graded soil would contain small particles, like sand, and larger particles, like gravel. Well-graded material can be well compacted. The aggregate used for roads is well-graded crushed sand and gravel. It can be well compacted because the particles "fit together"— the small particles fill the voids between the larger particles— which provides a better road base. Well-sorted material would include Ottawa sand which is used as filter pack around groundwater wells, or pea gravel which is often used when backfilling underground storage tanks. Each consists of a relatively narrow band of particle sizes. I also understand that golf courses order their sand for traps to be a uniform size, based upon sieve sizes, to make it more difficult to get out of the traps. My understanding is that they like a well-sorted sand, around the #50 sieve size.

To put this information to a practical environmental sampling use, the gradation of a soil will typically affect the porosity and permeability of a soil. For example, we use Ottawa sand as filter pack around a groundwater well because its uniform size allows free flow of water through it. It has a high porosity (approx. 30%) because it is well-sorted. If well-graded sand was used for a filter pack, the porosity would be lower so developing, purging, and sampling groundwater could take longer, as the well would not recharge as readily. Also, if the filter pack had a lot of fine-grained soil or fine sands in it, then they could move into the well and cause siltation in the bottom of the well. These concepts can be applied to in-situ soils, as well. The larger grain size soil will typically be associated with high porosity and high permeability.

Water in Geology

An aquifer is the place in the ground where you will find water in abundance that can be pumped out in economical useful volumes or rates — typically soil or rock that has high porosity and permeability. For the purposes of environmental sampling, you will not likely be conducting economical analyses during your sampling events. You should be more concerned with whether or not there is water and from what geological formation it is flowing. In a broad sweeping generality, water can be found in pumpable abundance in the following soils: gravel, sand, and silt. It can be abundant in clay, but you likely will not be able to pump water from a clay formation. In bedrock, abundant water can generally be found in highly fractured rock. However, if the rock is not fractured, there are only a few rock types that will support abundant water, for example, sandstone, limestone, and dolomite.

An aquitard is a layer of material that does not contain water in abundance and will stop the flow of water between geologic formations. Typical aquitard formations include clay, silty clay, and shale, each of which has a low permeability.

Field Assistance

When you are in the field, you don't have the luxury of a full service geotechnical lab to answer your questions about particle size. There are several tips, however, that can assist you in assessing soil particle size. For gravel, you simply have to measure it. If the particle is $^3/_8$ inch diameter or larger, it is gravel. The particle is sand if you can see a grain. You can describe it further as fine, medium, or course-grained sand. As you are unlikely able to measure the grains in the field, you should judge the size relative to the other grains. The important thing is to maintain consistency. If you call a certain size a medium-grained sand, make sure you call the same material by the same name in subsequent samples and borings.

For silt and clay, because you cannot see the particles with the naked eye, you have to use other methods to identify them. The first thing to do cut the top 1/8 inch along the length of the sample with a knife, (see The Ideal Tool Kit) assuming the sample came from a split spoon or other sub-surface sampling method. If the knife cut produces a shiny, smooth surface, the material is predominantly clay. If there are areas that have a layered, dull appearance, then those areas are likely silt. Further, take a small piece of the material (moisten if it is dry) and roll it out in your hand to make a long thread. If the material rolls into a thread with a thickness of approximately 1/8 inch, then it is mostly clay. If the material crumbles as you try to roll it, then it contains more silt than clay. This test is akin to one part of the Atterberg Limits, also known as Liquid Limit, Plastic Limit, and Plasticity Index of Soils ASTM 4318. The Atterberg Limits test was developed to determine the moisture content of clay when it becomes liquid, meaning it can flow (Liquid Limit). Also, it determines the moisture content of clay when it crumbles (Plastic Limit). Finally, if you suspect the material is all silt, put some in the palm of your hand and add a little water. If it is silt, the water will absorb into the material and "disappear." If you then gently tap the top of the silt with your fingers, or tap the hand that holds the material, the water will come to the surface of the sample. This is the "dilatency" test. Capillary action causes the water to rise when the silt is shaken gently. If the water stays "in" the material, it is likely fine sand.

> Career Tip – When you have nothing to do, go to your company's soils lab. Spend as much time there helping as you can. Get involved by preparing samples, running the tests, and logging the results. You'll understand more in the field after you spend time in the lab.

Soil Description

What you may see as silty clay, I may think is clayey sand. What I may describe as 'green,' you may describe as 'olive.' There are a few correct ways to describe soil, but there are many wrong ways. The Unified Soil Classification System (USCS) (ASTM D2487) is the standard system used to describe soil, while the Munsell® chart is the standard system for identifying colors. However, the USCS is, for engineering purposes and laboratory testing, required to accurately classify soil. It does have a "sister" standard practice called "Description and Identification of Soils (visual-Manual Procedure) (ASTM D 2488), which relies on field tests rather that laboratory testing. However, it is my opinion that if people are to use either of these standards, they should have both a good and practical understanding of the laboratory tests used in D2487. I will provide a better understanding of the field tests used in those practices.

Another method commonly used for soil classification is the United States Department of Agriculture's (USDA) Textural Classification Chart. This method uses the percent of sand, silt, and clay in a sample to determine its texture. The percentages of each constituent are plotted on a triangle (see *http://soils.usda.gov/procedures/handbook/content/618ex8.htm*) that has clay, sand, and silt on its sides. The triangle is subdivided based upon percentages of each soil type. The subdivisions are labeled with terms based upon these percentages. It is from this triangle that we get the term loam. Some regulatory agencies require the use of this USDA system for describing soil samples.

Soil Terminology

The exact methodology for the USCS can be found in ASTM D2487. In general, it takes the percentage of constituents, or particles, in a soil sample (based upon sieve analysis) and designates a two-letter symbol for that material. It is difficult to provide a full soils class in one chapter with no practical hands-on assistance, so I will provide the information that is helpful for completing a borehole log so that others reading it can understand what you are trying to convey. Keep in mind that various geologists and engineers may have their own nuances in the descriptions. That is their prerogative and, if that geologist or engineer is your boss, it becomes your prerogative. There is an order to follow when describing a soil: consistency, color, major soil component, adjective, minor soil component, USCS designation, then moisture content. For example: stiff gray silty clay, trace gravel. Each component is described further below.

1. Consistency/Relative Density

   This term refers to a soil's strength or bearing capacity, and there are different modifiers for cohesive and non-cohesive soils. First, cohesive soils like silt and clay have fines (passing the #200 sieve) in them and they stick together. Non-cohesive soils, like sand and gravel, do not.

   The bearing capacity, measured in tons per square foot (tsf), or, in metric, kilograms per cm^2 (kg/cm^2), refers to the strength of the soil. That is how much weight it can support. This is a geotechnical term that allows engineers to estimate the size of building and type of foundation that can be safely constructed on the soil. Relative density terms for gravel, sand, and silt (course-grained soil) include very loose, loose, compact, dense, and very dense. The term corresponds with the blow count (as shown in Table 1.1), which is obtained during drilling.

Cohesive soils (clay, silty clay, and clayey silt) are termed very soft, soft, firm, stiff, very stiff, or hard. These terms can be based upon the "N" value (blow count) or response to finger pressure (see Table 1.1). A stiff clay will support more than a soft clay. If your finger can push into the clay, imagine how much more so would a building.

The descriptors for bearing capacity are as follows:

Table 1.1. Soil Strength.

Cohesive Soils (Clays)				Non-Cohesive Soils (Sand, Gravel, Silt)	
Consistency	**N-value**	**Field Test**	**Pen. (tsf)**	**Compaction**	**N-value**
Hard	>30	Difficult to Indent	4	Very Dense	<50
Very stiff	15-30	Indented by Thumbnail	2.00 to 3.99	Dense	30-49
Stiff	8-15	Indented by Thumb	1.00 to 1.99	Medium Dense	10-29
Medium	4-8	Molded by Strong Pressue	0.50 to 0.99	Loose	4-9
Soft	2-4	Molded by Slight Pressure	0.26 to 0.49	Very Loose	0-3
Very Soft	<2	Extrudes between fingers	0.0 to 0.25		

2. Color

There is a device called a Munsell® chart that has on it all the colors of the rainbow, and every hue and shade in between. The chart was developed with the U.S. Soil Conservation Service to classify soil colors. You can purchase a binder (see *www.munsell.com* for ordering information) that has a full set of colors and a clear plastic sheet overlay for each page. The binder has 11 pages and almost 400 color chips on those pages, each with a name and an alpha-numeric reference.

The purpose of the binder is to help provide a consistency between samplers. When collecting a sample, take a small piece of soil and compare it to the colors in the binder by wiping the soil on the plastic overlay until you find the "match." This avoids various samplers describing the same soil in incongruous ways. Consistency is always important, but on larger jobs where four samplers are manning four drill rigs on 40 acres, they are all likely encountering the same geology thus making consistency essential.

Not everyone is required to use the Munsell chart, so the important thing when you are describing soil without a color chart is to be consistent. As the day drags on, you have to maintain that the olive green layer is still olive green, even if it appears different due to factors such as different levels of sunlight. If you change your description of the same sample, the person reviewing the log will think there is a geological or chemical change.

There are a few more points that need to be mentioned regarding coloring. Soil is typically a shade of brown or gray. It will usually be brown near the surface and become gray as you get deeper. The Munsell chart has an entire sheet devoted to the color gray, so be specific in your description. Soil that has always been below the water table, i.e., in an oxygen-poor environment, will be a shade of gray.

Mottling is a term that is often used to modify or enhance coloring descriptions. A mottled appearance will include different colors or shades. If you think of your skin after a long workout, it could have a blotchy red and white appearance. It is mottled red and white. With soil, you may have brown and gray mottling, where the two colors are among one another. You could have a light gray and a dark gray mottled together. This appearance in soil suggests a transitional area of the geology. You will often find brown with gray mottling as you approach the water table. As you get closer it may transition to gray with brown mottling, then to gray alone. When describing the soil, put the predominant color first then the secondary color.

Soil is usually brown or gray, but occasionally you may encounter green soil. Green soil does not tend to occur naturally, except near olivine or serpentine bedrock. When soil is impacted with petroleum products such as gasoline or diesel, you will find variations of green. So, if you find green soil while drilling near a potential source of petroleum, it is possible that the soil has been impacted. This does not mean the soil is contaminated, however. I encountered green soil when drilling within several feet of a former petroleum underground storage tank (UST), but the analytical results did not show contamination (based upon Risk Based Corrective Action Objectives–See Project Management).The soil was impacted by the petroleum, but not contaminated. The contact with petroleum likely occurred decades before, and the principal contaminants had probably volatilized, degraded, or migrated, leaving behind green soil.

Black soil does not tend to be abundant, except perhaps in Hawaii and the other volcanic regions. It occurs in some areas in the western United States containing a certain limestone. Regardless, you may encounter black soil. Based on my experience, black soil may indicate contamination caused by thick heating oil or motor oil/waste oil, or by metals in the soil. In the Chicago area, soil containing slag is often encountered.

It was used as fill from the multitude of steel mills in the vicinity, and the soil containing it is often black in appearance. Also, a rich, organic topsoil and peat will be black.

3. Major Soil Component

   The major soil component is that portion of the sample comprising at least 50% of the total weight. There are only 10 generally accepted terms when describing soil: gravel, gravel and sand, sand, clay, silty clay, clayey silt, silt, organic silt, and organic clay. Although many people use the term "sand and gravel," it is not present on the list because it is not technically acceptable. When writing the major component on a borehole log, it should be in upper case letters and/or be underlined. Major soil components can only be described accurately if you have an understanding of soil size (as presented in Section 2.2).

4. Adjective

   The adjective indicates the percentage by weight of the minor soil component. The four adjectives are: "and" (30-49% by weight), "some" (12-30%), "little" (5-12%), and "trace" (1-5%). I have seen other values for these adjectives and several other terms. For example, I have seen trace defined as 1-10% but it is still the adjective for the smallest value. As stated, the description is based on weight, not volume. If there is a one-inch seam of gravel in a two-foot sample of clayey silt, the volume is approximately 4 percent or "trace gravel." However, the weight might be closer to 7 percent, so it would be "little gravel." In ASTM D2488 (Soil Identification), there is a note saying terms may be used as follows: trace (<5%), few (5-10%), little (15-25%), some (30-45%), and mostly (50-100%). The key is to use the terminology of the senior personnel in your office or company, as they will be interpreting your logs and descriptions and will want you to use the terms with which they are familiar.

5. Minor Soil Component

   The minor component comprises less than 50% of the sample by weight. The allowable major soil components are used for the minor components as well. You should not describe minor silt and clay portions separately. For example, "little silt, trace clay" should be "little clayey silt".

6. USCS Symbol

   This is a two-letter symbol provided from ASTM D 2487.

7. Moisture Content

   Moisture Content describes how much water is in the soil using the terms saturated, wet, moist, damp, or dry. When entering this modifier on a borehole log, it usually is written in a separate column, not with the rest of the sentence.

So, an acceptable description found on a borehole log would be, "Compact, gray f-m SAND, some Silt (SM). Wet" The f-m before sand designates fine to medium. Another example would be, "Stiff, tan, SILTY CLAY, little fine sand (CL). Moist."

A tip to remember is to not mix the consistencies between cohesive and non-cohesive soils, i.e. clay is not dense and sand is not soft.

Those are the basics for soil identification. There are other descriptions from ASTM D 2488, a few of which I have described below. The U.S. Department of Agriculture has adopted this standard for soil identification, as have certain states. The thing to remember with ASTM D 2487 and D 2488 is they are for engineering purposes. It is one thing to know the soil if you are building a bridge or a 50-story tower on it it is another if you are collecting samples to determine contamination then dig and haul it out to a landfill. Keep in mind, however, if you are encountering boulders during your investigation, either suspected in drilling, or observed in test pits, make sure you note that in your log. The cost of excavation increases when boulders are encountered.

Additional characteristics to consider during environmental sampling include:

1. Particle angularity, described with the following terms: angular, subangular, subrounded, and rounded. These terms are used to describe the shapes of gravel or course sand in a sample. From an environmental standpoint, sand and gravel that is angular will be better compacted than rounded particles. The rounded particles would then have a higher porosity and likely a higher permeability (i.e., water would flow through round particles quicker than angular).

2. Particle shape, described as flat, elongated, or flat and elongated. Particle shape is only to be mentioned for gravel and cobbles, and only if they are flat and/or elongated.

3. Maximum particle size or dimension.

4. Odor, which is mentioned only if organic or unusual. You do not have to explain throughout the borehole log that the sample had no unusual odor only when there is an odor other than a "soil-like" odor.

5. Reaction with HCl, described in the terms none, weak, strong. Limestone will effervesce when HCl is applied to it. Dolomite will not effervesce unless you scratch the surface to create a dolomite powder.

6. Structure should be noted for intact samples. Structure can be stratified, laminated, fissured, slickensided (polished or smooth, lensed, homogeneous.

7. Additional comments: presence of roots or root holes, presence of mica, gypsum, etc., surface coatings on coarse-grained particles, caving or sloughing of auger hole or trench sides, difficulty in auguring or excavating, etc. The presence of roots or other organics is an important feature with environmental sampling, as carbon (present in organics) will absorb contaminants or aid in limiting their migration potential.

Soil Permeability

From an environmental standpoint, the difference in the size of sand particles (fine, medium, or coarse) is very important. The grain size dictates the permeability and the porosity of the soil. Both of those are important to know how fast water and contaminants will flow through the soil. Particle size usually dictates where and how fast a contaminant will flow. In gravel, the permeability is high, so water and contaminants will flow quickly through gravel. The article of measure for permeability is cm/second. Generally, soil with a permeability of 10^{-6} cm/second (clay) is impermeable, 10^{-4} to 10^{-5} cm/second (fine sand, silt) is relatively permeable, and $< 10^{-4}$ cm/second (sand and gravel) is highly permeable.

Borehole Logs

In the beginning of your career, when in the field manning a drill rig one of your primary tasks is to provide somebody in the office with a description of the geology. The other primary task is to collect that geology and put it in sample containers—see Chapter 4. Later in your career, you too will rely heavily upon the field writings of some young one just out of college. The method to transfer this important information is via the borehole log.

The typical borehole log measures 8.5-by-11-inches—a standard sheet of paper. I recommend that you increase the size of your borehole log to 11-by-17-inches Project Managers should you insist on it. It provides more than twice as much room (101 more square inches) to write that crucial information. Among my faults in the field was poor writing. I would compound that issue by trying to write too much information in small areas, making it difficult for project managers to read; sometimes even I could not decipher my own efforts. You want to avoid having your project manager ask if you were referring to silty clay or a

Borehole Log

Project No. ____________ Job Name ____________ Date ____________ Borehole No. ____________
Borehole Location ____________ Elevation ____________
Type of Rig ____________ Contractor ____________ Driller ____________
Weather ____________ Temperature ____________ Logged by ____________

Time Start ______ Time Complete ______
Water Level Data

______ Ft. while drilling

______ Ft. at completion

______ Ft. Hrs after drilling

Cohesive Soils (Clays)				Non-Cohesive Soils	
Consistency/N-Value		Field Test	Pen. (tsf)	Compaction	N-value
Hard	>30	Difficult to Indent	4	Very Dense	<50
Very stiff	15-30	Indented by Thumbnail	2.00 to 3.99	Dense	30-49
Stiff	8-15	Indented by Thumb	1.00 to 1.99	Medium Dense	10-29
Medium	4-8	Molded by Strong Pressure	0.50 to 0.99	Loose	4-9
Soft	2-4	Molded by Slight Pressure	0.26 to 0.49	Very Loose	0-3
Very Soft	<2	Extrudes between Fingers	0.0 to 0.25		

	SOIL STRATIGRAPHY		SAMPLES				DESCRIPTION AND BORING NOTES
Depth	Description	"N" value	Moisture	Recov.	No.	PID	

Special Notes:

Project No. ____________
Borehole No. ____________
Depth ________ to ________

Figure 1.6. A borehole log.

sandy clump, or to green streaks or gray streaks? It is also a good idea to transfer these logs to the computer as soon as you get back from the field.

The 11-by-17-inch format not only provides more space to describe the soil that is being sampled, but it also accommodates notes to assist you in that description (see Figure 1.6).

The attached example Borehole Log was created in Excel and fits nicely onto an 11-by-17-inch format. It includes sufficient space to describe the sample and other notes. The following provides additional information regarding completing the log.

If you use this larger format, you first have to make a clipboard for it. I cut a piece of ¼-inch plywood, good on one side, which measured 12-by-18-inches I used four binder clips to attach the paper to the board , one on each side. Some others with whom I worked, who were more adept at such things, riveted an actual "clipboard" device to the top of the wood, making for a more permanent device. Since this clipboard cannot be closed, you can get a piece of plastic overlay to protect your log from rain and mud splattering. You can also permanently affix various logging notes (e.g., USCS charts) to the plywood and simply lift the borehole log to reference the information.

Going through the various entries at the top of the borehole log, it seems quite evident what you should fill in. I would like to emphasize a few of them, however. The borehole location, and subsequent well location, is very important. The more precise data you can provide will save hours of time in the future when a colleague (or you) cannot find the well, especially flush-mount wells. Flush mounts are particularly difficult to find under snow and I know of one that was covered with sod. When you measure a location, use permanent fixtures as your reference. Do not use trees, as they can be cut down and others can be planted. Do not use sign posts because they can be run over. Try to use the sides of buildings, unless you know a demolition is slated or use sidewalks or road curbs. Make sure you reference the side of the sidewalk, e.g., 127 feet north of north edge of sidewalk. Because sidewalks are three feet wide just saying "north of the sidewalk" will only provide a range. Other fairly permanent features include fire hydrants, rail lines (mention which track), manholes, and well marked property lines. Less permanent features include the edge of parking lots (they get expanded and re-paved), guard shacks, which can be moved, and survey stakes, which are always targets of equipment operators.

Also, many people will put survey tape around a well, especially in trees surrounding a well, to mark its location. As they tie the red, pink, or yellow ribbons in the lush green trees, it is evident to them where the well is. Three months later, as the leaves are turning red, pink, and yellow, the survey tape does little to demarcate a well location. Also, this tape fades in sunlight, so what was brilliant when attached is dull when needed. While the tape serves its purpose in the immediate, you cannot rely on it when you send someone out in the woods to find a monitoring well installed months or years earlier. You must rely on good measurements. If you are out in the woods, there will not be any of the primary reference points mentioned in the paragraph above. In this case, consider a global positioning system (GPS). If the budget does not provide for this, (although they are getting quite inexpensive), try the trusted compass and pacing factor.

The Type of Rig, Contractor, and Driller are quite evident. If you do not know the type of rig, ask the driller. Weather, Temperature, and Logged by (you) are self-explanatory. Make sure you get the start and completion times of drilling. If you are installing a well, make sure you note the time that activity is started and ended. Some drillers charge a different rate when not drilling. It is up to you to track their time.

For Water Levels, if water is encountered during drilling, meaning samples are coming out saturated, have them stop briefly to get the water level. If the auger is sticking out of the ground, assuming hollow stem augers, measure to the top of the auger, grasp that point on the tape and pull it down to the ground. Then, look at the measurement at the top of the auger. That is your depth to water. That saves doing math with feet and inches. When they have finished drilling, get another water level. Then, if the borehole is still open after clean up, etc., grab another level. If the hole caved in and you are unable to measure water, note to what depth it caved. This is good information, should you be planning an excavation. If you are installing a monitoring well, get the water level upon completion. If you use the same water level monitor throughout, make sure you decontaminate it properly before putting it into the monitoring well. Finally, if there is no water in the boring or well, write NWE, which stands for No Water Encountered, not "dry." There is moisture in virtually everything, especially soil at depth.

Moving down on the borehole log, Depth should be self-evident, although you can establish the scale to your liking. One or two lines per foot would fit nicely. In the Description column, you should put the soil description as discussed above. This is where you write, for example, "stiff gray <u>SILTY CLAY</u>, trace sand." You should put the major component in caps and/or underline it. For the "N" Value column, write the number of blows per six inches that the two-inch diameter split spoon is driven. Some drillers count that for you, others do not , so you should find out who is counting before you start drilling. The "N" value is the sum of the either the last two or middle two six-inch counts. If the split spoon is 18 inches long, use the last two numbers. If it is a two-foot spoon, use the middle two numbers. You never use the first number because it could represent the "sluff" (cuttings) that has fallen to the bottom of the boring. Write the counts for each six inches in the column, total the two numbers you are using, write the sum, and circle it. The circled value is the "N" value. It is used to assess soil strength and, as stated above, it determines the density/consistency aspect of your soil description.

Next we have Moisture, described as damp, moist, wet, or saturated. Again, avoid using "dry" unless it is powdery or dusty. Moving across, there is Recovery. I usually put that as a fraction of the total length of the sampling tool. If you are using a two-foot spoon, and 12 inches of soil is in the spoon, the recovery is 12/24. The recovery is simply the measurement of the soil retained in the sampling tool. If you are using a GeoProbe®, then your recovery would be over 48 inches. So, if there is 42 inches of soil in the tube, the recovery is 42/48. If you only drive the sampler a portion of its length, do not write down its total length as the denominator; use the actual length driven. Recovery is more of a geotechnical term, but the information can be useful for excavations. Also, the recovery will diminish in wet or saturated non-cohesive soil zones.

The Sample Number is the sample number. The PID reading can be obtained one of two ways. First, and most likely, is the headspace measurement. Others might pass the PID over the length of the sample and note the readings and peaks. If that is the method you use, you still have to write the headspace reading somewhere on the log. That could be under the Description and Boring Notes column.

The Description and Boring Notes will include virtually everything else. Here you can put the Munsell® color designation number. If you use HCl, mention here if it effervesces. If you use a penetrometer, write the reading in this area. You can describe further the sample that you collect. List the specific location, e.g., sample collected from 3.5' to 4.0'. You can mention if there is an odor associated with the sample—remember, no odor, don't mention it. Other attributes of the sample can be listed here, e.g., iridescence, product in sample, one-inch stone at 2.7 feet, etc. You can mention driller's comments in this area. He may tell you that the borehole has caved in at a certain depth, if the geology or soil density has changed (sometimes evident to a driller), if he has hit a cobble or boulder, etc. If you take water levels throughout boring, write those under the notes.

For the Special Notes area, you can include additional groundwater conditions, hours of productive drilling versus hours delayed, reasons for delay,

Finally, when you write your boring log, write neatly and try to keep mud off of it. Some people write in their field books and transfer it to a log in the office under ideal conditions, or they do one in the field and redo it in the office. I do not think that time should be charged to the client. You are in the field to write your field notes, so do it there. If you have poor handwriting, write slowly, take your time, and use a straightedge to draw lines. Many decisions regarding remedial options and design could be based upon your borehole logs, so make them as accurate and complete as possible.

2

Chemistry Fundamentals

The reader should possess a basic understanding of some fundamentals of chemistry in order to fully comprehend this chapter. That understanding should include recognition of a periodic table, some basic chemical formulae, and knowledge of the fact that the primary reason to collect samples is to assess the presence of chemicals (or elements/ions) in the sample media.

Chemical Terms

Each chemical known has a number attributed to it called the CAS number. CAS is the Chemical Abstract Service, located in Columbus, Ohio. For nearly a century, this group has provided this service to avoid confusion caused by the various names by which chemicals are called. For example, trichloroethylene, trichloroethene, and TCE are the same chemical, and its CAS No. is 79-01-6. 1,1,1-trichloroethane is synonymous with methyl chloroform and its CAS No. is 71-55-6. Also, trichloromethane is also known as chloroform, with the CAS No. 67-66-3. To the chemical lay person, trichloroethene, 1,1,1-trichloroethane, and trichloromethane sound similar, if not the same. However, they differ greatly in their appearance, use, and toxicology. Therefore, familiarity with CAS numbers is important to avoid chemical confusion.

The use of CAS numbers is important because if your analytical results come back positive with 1,1,1-trichloroethane and the facility you are investigating has never used that chemical, you might question how it got onto the property. However, if a search of the facility's records indicates having used methyl chloroform or CAS No. 71-55-6, then you have a logical source, and you can pursue the source and the plume with added knowledge. There are over 43 million numbers in the system, so you are not expected to memorize them all. However, you should be aware of synonyms (see Table 2.1) and be able to use the CAS numbers to clarify the chemical with which you are concerned.

Another scenario where the CAS numbers are useful is when ordering Material Safety Data Safety Sheets (MSDSs). When preparing a Health and Safety Plan, you will often have to order MSDSs. It is crucial that you order the proper sheet, and using a chemical's CAS number will assist in that endeavor.

The following sections discuss chemicals for which we routinely sample. I have summarized these chemicals in tables and have included their synonyms and CAS numbers. It is often frustrating when sampling for these compounds to be unsure of what they are or why they are used. I have included descriptions of the chemicals, how they are used, and where one would expect to encounter each chemical in order to ease this frustration.

The typical suite of chemicals for which one samples includes volatile organic compounds (VOCs), semi-volatile organic compounds (SVOCs), polynuclear aromatic hydrocarbons (PAHs), base/neutrals/acids (BNAs), total petroleum hydrocarbons (TPHs), metals, pesticides, herbicides, and polychlorinated biphenyls (PCBs). In addition, you may also collect samples to test water quality parameters. These may include biological oxygen demand (BOD), total dissolved solids (TDS), total suspended solids (TSS), chemical oxygen demand (COD), dissolved oxygen (DO), and pH. The chemicals tested for can be separated into three groups; organics, inorganics, and wet chemistry. This is often how an analytical laboratory will divide its equipment and employees. I'll explain what these parameters are, why we analyze for them, and other characteristics of the chemicals that should help you understand more about your sampling efforts.

Organic Chemistry

There are exceptions, but generally, organic chemistry is the study of chemical compounds that contain carbon. Some characteristics of typical organic compounds include the following:

1. They are flammable,
2. They have low boiling and melting points,
3. They are insoluble in water, and
4. They are soluble in organic solvents (which is why trichloroethylene is a good degreaser).

Taking a class in organic chemistry will help you better understand your environmental work. The class is generally difficult, so you may want to take it at night, perhaps at a community college, in order to keep that grade separate from those of your other classes. This will enable you to concentrate on learning the subject matter rather than your grade. For the purposes of environmental sampling, organic chemistry encompasses volatile organic compounds, semi-volatile organic compounds, petroleum products, pesticides, herbicides, and PCBs.

Volatile Organic Compounds

In the simplest of terms, set by the EPA, VOCs are those carbon-containing chemicals that evaporate at room temperature. More technically, VOCs are defined based upon their vapor pressure, which is a chemical's tendency to evaporate. It is measured in mm of Hg, torrs, psi, and atmospheres (1mm of Hg = 1 torr, 760 mm of Hg = 1atm = 14.7 psi). In Australia, a chemical is a VOC if it has a vapor pressure > 2mm Hg at 25°C and contains carbon. The EPA also defines a VOC regarding its propensity to photochemical reactivity, meaning it will change when exposed to light. There are some VOCs that evaporate quickly at room temperature but do not readily photo-react, therefore, they are not VOCs with which the EPA has a concern. Another source, SW-846 Method 8260B, Section 1.3, states that if a chemical has a boiling point less than 200°C, it is a VOC. Method 8260B lists numerous chemicals as VOCs; however, the list is long and includes chemicals for which few would analyze. There are several sources for the lists of compounds for which one must analyze. The EPA has a Target Compound List (TCL) as part of their Superfund division and Contract Laboratory Program (CLP). Labs that work with federal environmental programs (e.g., Superfund and Federal Brownfields) must adhere to the many guidelines issued by CLP. These requirements are complex and do not often reflect activities in the "real world."

A list of VOCs for which one commonly samples and analyzes is presented in Table 2.1. As discussed above, the table also lists synonyms for chemicals, CAS numbers, and the health hazards associated with each chemical. Do not use this list as a default for your work, it is only a guideline. You, your client, and the regulating authority should define the actual list you use. When defined, that list has to be provided to the laboratory so they analyze what you want. For all analyses, you must specify your definition of a VOC, as the lab's standard VOC list may reflect a list designated by the state in which they operate.

Many of the chlorinated contaminants listed in Table 2.1 may be present at a site, even if they have never been used there. Select chlorinated solvents break down or degrade into other chemicals over time and distance from their source. Specifically, 1,1,1-trichloroethane (1,1,1-TCA), a degreaser, and perchloroethylene (PCE), a degreaser and dry cleaning fluid, break down to eventually become vinyl chloride (VC) and chloroethane, with several daughter products in between. This occurs primarily in anaerobic environments. Vinyl chloride is a known human carcinogen and chloroethane has caused cancer in mice. There are several other chemicals in between TCE and VC, as shown in Figure 2.1. So, if you are collecting samples at a site where TCE was used extensively (e.g., a facility that used degreaser), you will have to analyze not only for the presence of TCE, but for daughter products as well, including vinyl chloride.

Table 2.1. Volatile Organic Compounds

Compound	Synonyms	CAS No.	Use	Toxicology (Why you're looking for it)
Acetone	Dimethylketone, 2-Propoanone, pyroacetic ether	67-64-1	Widely used as a solvent for fats, oils, waxes, and rubber cement. Used in manufacture of chloroform explosives, rayon, photographic films. Used to store acetylene gas.	Irritation to eyes, nose, and throat (ENT), headaches, dizziness and dermatitis. Affects Kidney.
Benzene	Benzol, phenyl hydride, coal naphtha, benxole, cyclohexatiene	71-43-2	Constituent of motor fuels (gasoline), solvent for fats, inks, oils, paints, plastics, and rubber. Chemicals such as styrene and phenols are produced from benzene and it is used in manufacture of detergents, explosives, and pharmaceuticals.	Carcinogen and other nasty effects.
Bromodichloro-methane	Dichlorobromomethane	75-27-4	Used in organic synthesis	Carcinogen to mice and rats. Irritates eyes, nose, mucous membranes. Affects kidney.
Bromoform	Tribromomethane, methyl tribromide	75-25-2	Used in pharmaceutical manufaccturing, fire-resistant chemicals, and guage fluid. Solvent for waxes, greases, and oils.	Irritation to ENT, headaches, dizziness and death. Carcinogen to lab animals. Affects gastointestianl system.
Bromomethane	Methyl bromide	74-83-9		Affects respitory system.
2-Butanone	Methyl ethyl ketone, butanone, MEK	78-93-3	Solvent in nitrocellulose coating and manufacture of vinyl film, smokeless powder, cements, adhesives and in dewaxing of lubricating oils. Used in drug manufacture.	From irritation of ENT to death.
Carbon Disulfide	Carbon bisulfide	75-15-0	Manufacture of rayon, ammonium salts, optical glass, paints, paint removers, varnishes, explosives, rocket fuel, preservatives, and rubber cement. Used as solvent and deagreaser. Also used in electroplating, grain fumigation, and dry-cleaning.	Horrible stuff. Various and sundry effects including reproductive system.
Carbon Tetrachloride	Tetrachloromethane, percholormethane	56-23-5	Solvent for oils, fats, lacquers, varnishes, waxes and resins. Used as a dry-cleaning agent, fire extinguishing agent, and fumigant.	Attacks various organs, causes death.
Chlorobenzene	Monochlorobenzien, chlorobenzol, phenyl chloride, MCB	108-90-7	Used in manufacure of aniine, phenol, chloronitrobenzene, and many pesticides.	Burns skin, irritate ENT, organ damage.
Chloroethane		75-00-3	Formerly used in gasoline, used to numb skin prior to piercings, biopsies. Used as in drug mfg., as a solvent and refrigerant.	Affects nervous system causing loss of muscle control, unconsciousness.

Table 2.1. Volatile Organic Compounds (Continued)

Compound	Synonyms	CAS No.	Use	Toxicology (Why you're looking for it)
Chloroform	Trichloromethane, methane trichloride.	67-66-3	An early anesthetic but use stopped due to its toxic effects. Used as solvent, used in pharmaceuticals, manufacture of plastics, floor polishes and fluorocarbons, and catgut sterilization.	Fainting, vomiting, dizziness, coma, organ damage, suspected human carcinogen, death.
Chloromethane	Methyl Chloride	74-87-3	Found in polystyrene insulation, cigarette smoke, aerosol propellants, chlorinated swimming pools. Produced in nature in oceans. Produced when grass, leaves, and charcoal are burned.	Affects nervous system, including convulsions and coma. Also affects your liver, kidneys, and heart.
Dibromochloro-methane	Chlorodibromomethane, CDBM	124-48-1	Used in manufacture of fire extinquishing agents, aerosol propellants, refrigerants, and pesticides. Often found in drinking water, may be formed during chlorination.	Irritant, narcotic, organ damage.
1,1-Dichloroethane	Ethylidene chloride	75-34-3	Used as solvent and cleaning and degreasing agent.	Fetus damage, organ damage, death.
1,2-Dichloroethane	1,2-Dichloroethylene, ethylene dichloride, ethylene chloride.	107-06-2	Used in manufacture of ethylene glycol (antifreeze), PVC, rayon, nylon, and plastics. Solvent for resins, asphalt, bitumen, rubber. Extracting agent for soybean oil and caffiene. Used as anti-knocking agent in gasoline, a pickling agent, fumigant and drycleaning agent. Used in photography, xerography, water softening and in production of adheisives, cosmentics, pharmaceuticals, and varnishes.	From dizziness to death.
1,1-Dichloroethene	Vinylidene chloride, 1,1-dichloroethylene, DCE, VDC.	75-35-4	Used in manufacture of 1,1,1-trichloroethane and plastic food wrap and shrink wraps. Used as admixture in concrete and paints. Used to produce fibers for automotive uphostery, drapery, venetian blinds, and outdoor furniture.	Carcinogen, organ damage, irritates eyes and skin.
cis-1,2-Dichloroethene	Acetylene dichloride, 1,2-dichloroethylene.	156-59-2	Solvent for waxes and resins. Used as refrigerant, in manufacture of pharmaceuticals and artificial pearls, and in extraction of oils and fats from fish and meats.	Affects circulatory system.
trans-1,2-Dichloroethene	Acetylene dichloride, 1,2-dichloroethylene.	156-60-5	Solvent for waxes and resins. Used as refrigerant, in manufacture of pharmaceuticals and artificial pearls, and in extraction of oils and fats from fish and meats.	Affects circulatory system.

Table 2.1. Volatile Organic Compounds (Continued)

Compound	Synonyms	CAS No.	Use	Toxicology (Why you're looking for it)
1,2-Dichloropropane	Propylene Chloride	78-87-5	Used as a chemical intermediate in perchlorothylene and carbon tetrachloride synthesis, as a lead scavenger for anti-knock fluids. Solvent for fats, oils, waxes, gums, and resins. Also used as fumigant, metal degreasing agent, and insecticide.	Irritates ENT, cause dermatitis, rash, dizziness.
1,3-Dichloro-propene	Telone® 1,3-dichloropropylene	542-75-6	Used as soil fumigant. Applied to roots to kill nematodes.	Damages organs, gastrointestinal system. Causes headaches, affects ENT.
cis-1,3-Dichloro-propene	cis-1,3-Dichloropropylene	10061-01-5	Used as soil fumigant. Applied to roots to kill nematodes.	Damages organs, gastrointestinal system. Causes headaches, affects ENT.
trans-1,3-Dichloro-propene	trans-1,3-Diclhoropropylene	10061-02-6	Used as soil fumigant. Applied to roots to kill nematodes.	Damages organs, gastrointestinal system. Causes headaches, affects ENT.
Ethylbenzene	Ethylbenzol, phenylethane	100-41-4	Used in manufacture of styrene and synthetic rubber. Component of automotive and aviation gasoline.	Kidney, liver, skin, and respitory disease.
2-Hexanone	methyl n-butyl ketone, MBK, or propyl acetone.	591-78-6	Formed in manufacture of gas from coal, processing oil shale or wood pulp. Formerly used (pre-1982) in paint thinners and as solvent for oils and wax. No longer mfg. in USA.	Affects nervous system causing numbness, weakness and tingling in hands and feet.
4-Methyl-2-pentanone	Methyl Isobutyl Ketone, isobutyl methyl ketone, MIBK, hexone	108-10-1	Used in manufacture of methyl amyl alcohol and as a solvent in paints, varnishes, and lacquers	Dizziness, kidney and liver damage, irritant to ENT, skin, and mouth.
Methyl Tertiary-Butyl Ether	MTBE	1634-04-4	Used in gasoline. Also used to dissolve gallstones.	May affect nervous system. May cause nausea, throat and nose irritation.
Methylene Chloride	Dichloromethane, methylene dichloride.	75-09-2	Used as paint remover and degreaser. Used as solvent for oil, fats, waxes, and bitumen.	Effects range from loss of coordination to death. Animal carcinogen
Styrene	Cinnamene, phenethylene, styrol, styrolene, vinylbenzene	100-42-5	Used to form polystyrene and acrylonitile-butadiene-styrene (ABS).	Animal carcinogen. Dizziness to death.

Table 2.1. Volatile Organic Compounds (Continued)

Compound	Synonyms	CAS No.	Use	Toxicology (Why you're looking for it)
Tetrachloro-ethene	Tetrachloroethylene, PCE, perchloroethylene, carbon dichloride.	127-18-4	Used as dry cleaning agent, degreaser, and fumigant.	ENT irritant to death. Animal carcinogen
1,1,2,2-Tetrechloro-ethane	Acetylene tetrachloride, ethane tetrachloride	79-34-5	Used as dry cleaning agent, fumigant, and in cement and lacquers.	ENT irritant to death. Animal carcinogen
Toluene	Toluol, methylbenzene, phenylmethane, methylbenzol	108-88-3	Component of aviation and automotive gasoline.	Dizziness to death.
1,1,1-Trichloroethane	Methyl chloroform	71-55-6	Substitute for carbon tetrachloride. Used in degreasing and dry cleaning.	Dizziness to death.
1,1,2-Trichloroethane	Vinyl trichloride, beta-trichloroethane	79-00-5	Used as solvent.	Dizziness to death. Animal carcinogen.
Trichloroethene	Trichloroethylene, TCE, tri, ethylene trichloride	79-01-6	Used as solvent, as a caffeine extractant, in production of pesticides, waxes, gums, resins, tars, paints, and varnishes.	Dizziness to death. Animal carcinogen.
Vinyl Chloride	Chloroethylene, chloroethene, monochloroethylene	75-01-4	Used in manufacture of poly vinyl chloride (PVC) and other resins.	Human carcinogen
Xylenes (total)		1330-20-7	Used in paint, thinners, shellac. Found in gasoline and other petroleum products.	Affects central nervous system causing loss of balance, confusion, dizziness.

The daughter products and a degradation flow chart are presented in Figure 2.1. You will also analyze for VOCs and SVOCs (semi-volatile organic compounds) because the material being degreased is grease or oil products. And, because it is being degreased from pieces of metal you will also analyze for metals.

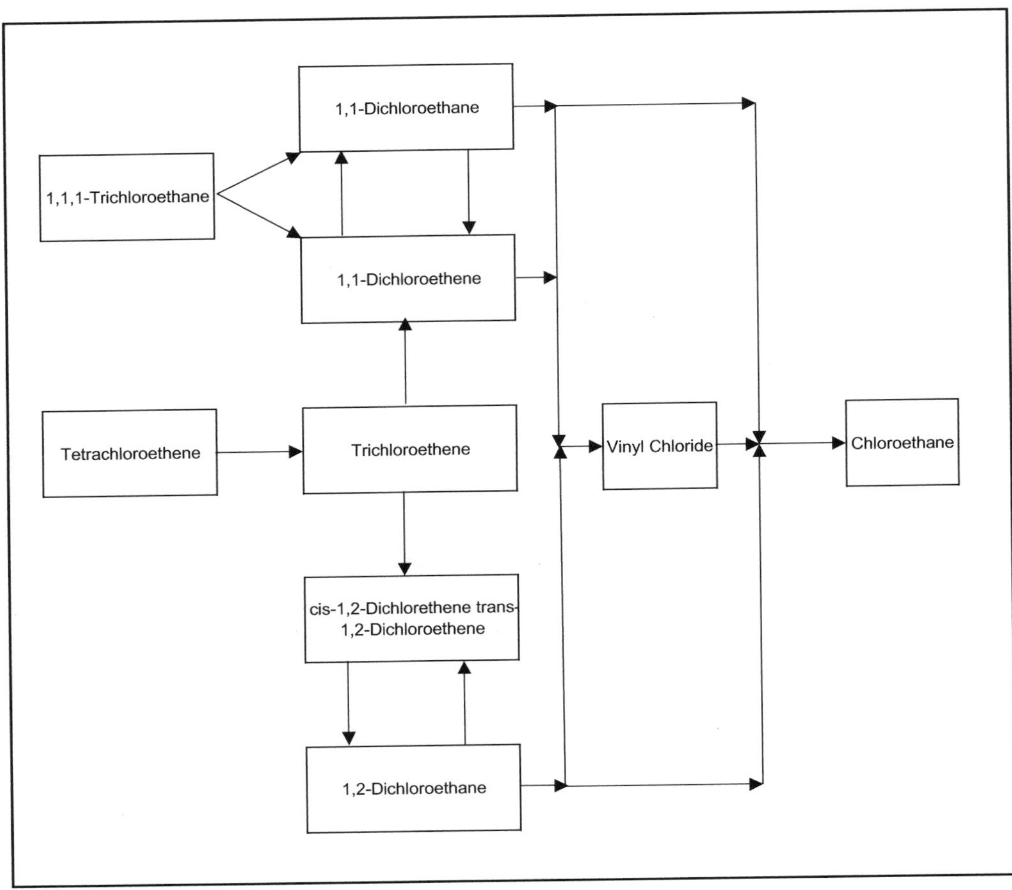

Figure 2.1. Biodegradation of Chlorinated Solvents

Semi-Volatile Organic Compounds

Semi-volatile organic compounds (SVOCs) are those compounds that are not as volatile as VOCs. The EPA defines SVOCs as chemicals with carbon that evaporate slowly at room temperature (20°C). You may also see SVOCs called semi-volatile organic analytes (SVOAs) or Base/Neutral/Acids (BNAs). BNA is a way of separating the compounds further and is based upon how the compounds are extracted during analysis. Base/Neutrals are extracted from a sample using a chemical with a pH > 7. Acids are extracted using a chemical with a pH < 7. However, keep in mind that BNAs are SVOCs. Table 2.2 lists SVOCs. Table 2.3 lists base/neutral compounds, and Table 2.4 lists acid compounds.

Table 2.2 Semi-Volatile Organic Compounds (SVOCs)

Compound	Synonyms	CAS No.	Use or Source	Toxicology (Why you're looking for it)
Acenaphthene	1,8-ethylenenaphthalene	83-32-9	Occurs in coal tar (produced by coking of coal) and used in manufacture of some plastics, as insecticide and fungicide.	Skin irritant and may cause vomitting.
Acenaphthylene	Cyclopenta(de)-naphthalene	208-96-8	Coal tar processing.	Causes irritation to digestive and respitory system, skin, eyes.
Anthracene	Paranaphthalene	120-12-7	Used in insecticides and wood preservatives.	May cause mutations. Throat and nose irritant.
Benzo(a)anthracene	Benzphenanthrene, tetraphene, benzanthrene, BA	56-55-3	Found in exhaust, soot, coal and gas works emissions, and cigarette smoke. Also found in waxes, creosote, coal tar, and petroleum asphalt. Found in charcoal broiled, barbecued, or smoked meats and fish and roasted coffee.	Animal carcinogen.
Benzo(a)pyrene	B(a)P (BAP), 3,4-benzpyrene	50-32-8	By-product of coal combustion. Also found in asphalt and cigarette smoke.	Animal carcinogen. Affects gastrointestinal system.
Benzo(b)fluor-anthene	B (b) F, benz(e)-acephenanthrylene	205-99-2	Formed during incomplete combustion of organic matter (fossil fuel, garbage). Component of creosote, coal tar pitch, and bituments.	Animal carcinogen.
Benzo(g,h,i)-perylene		191-24-2	A PAH caused by incomplete combustion of coal, meat, oil, gas, garbage. Used in mfg. of explosives, plastics, pesticides, drugs. Used to make steroids, cholesterol, and bile acids.	No known harmful affects in humans.
Benzo(k)-fluoranthene	Dibenzo(b,jk)fluorene	207-08-9	By-product of coal combustion. Also found in asphalt and cigarette smoke.	Affects gastrointestinal system. Animal carcinogen.
bis(2-Chloroethoxy) methane	BCEXM, dichlorethyl formal	111-91-1	Used in treatment of textiles, manufacture of polymers, insecticides, and as degreasing agents.	Toxicological properties have not been evaluated fully.
bis(2-Chloroethyl) ether	Dichloroethyl ether, dichloroether, BCEE, dichlorethyl oxide.	111-44-4	Used in manufacture of paint, varnish, lacquer,soap, and finish remover. Solvent for fats, oils, naphthalenes, greases, pectin, tar, and gum.	Animal carcinogen, coughing, tearing, retching, burns to skin and eyes, death.
bis(2-Ethylhexyl) phthalate	Di (2-ethylhexyl) phthalate	117-81-7	Used as a plasticizer for resins.	Animal carcinogen.

Table 2.2 Semi-Volatile Organic Compounds (SVOCs) (Continued)

Compound	Synonyms	CAS No.	Use or Source	Toxicology (Why you're looking for it)
4-Bromophenyl-phenyl ether		101-55-3	Has been found in raw water, dringing water, and in river water.	Skin contact leads to dermatitis.
Butylbenzylph-thalate		85-68-7	Used as a plasticizer for polyvinyl resins.	Skin irritant, carcinogen.
Carbazole	9-Azafluorene, dibenzopyrrole	86-74-8	Used in dyes, explosives.	Irritant to skin, eyes, digestive system.
4-Chloroaniline	para-chloroaniline	106-47-8	Used to make dyes, other chemicals, insecticides and various industrial products	Headaches, trouble breathing, weakness, bluish color ot nose and lips, collapse, and death.
4-Chloro-3-methylphenol	4-chloro-m-cresol	59-50-7	Germicide, preservative for glues, gums, paints, inks, textile & leather goods.	Can harm eyes, skin, toxic to fish.
2-Chloronaphthalene		91-58-7	Wide use in industry - including in production of electric condensers, insulation of electric cables and wires, additives to pressure lubricants.	Causes chloracne (pustules, papules, cysts).
2-Chlorophenol		95-57-8	Used in manufacture of fungicides, pesticides, herbicides, disinfectants, and wood and glue preservatives.	Severe irritation and burns to nose and throat, headache, dizziness, lung damage.
4-Chlorophenyl-phenyl ether	Benzene, 1-chloro-4-phenoxy	7005-72-3	Used as a dielectric fluid	May damage digestive track. May cause irritation to eyes and skin.
2,2'-oxybis (1-chloropropane)	bis(2-Chloroisopropyl) ether bis(2-Chloro-1-methylethyl) ether	108-60-1	Used in manufacture of dyes, resins, pharmaceuticals, and textiles. Also used at fungicide and insecticide in wood preservatives. Used in mfg. of glycols.	Inhalation, ingestion, or skin contact may cause severe injury or death.
Chrysene	1,2-Benzophenanthrene	218-01-9	Found in gasoline and diesel exhaust, cigarette smoke, and coal tar.	Animal carcinogen, skin rashes, pigment change.
Dibenz(a,h)-anthracene	DBA, 1,2,3,6-dibenzanthracene	53-70-3	Formed during incomplete combustion of organic matter (fossil fuel, garbage). Component of creosote, coal tar pitch, and bituments.	Animal carcinogen.
Dibenzofuran	2,2'-Biphenylen oxide	132-64-9	Used as an insecticide	Toxic but details not known.

Table 2.2 Semi-Volatile Organic Compounds (SVOCs) (Continued)

Compound	Synonyms	CAS No.	Use or Source	Toxicology (Why you're looking for it)
1,2-Dichlorobenzene	1,2-DCB	95-50-1	Used as solvent and in manufacture of dyes, hericides and degreasers.	Blood cell damage, skin and eye irritant, severe burns, damages liver, kidneys, and lungs
1,3-Dichlorobenzene	1,3-DCB	541-73-1	Contaminant of 1,2 and 1,4-DCB formulation	Blood cell damage, skin and eye irritant, severe burns, damages liver, kidneys, and lungs
1,4-Dichlorobenzene	1,4-DCB	106-46-7	Air deodorant and insecticide.	Blood cell damage, skin and eye irritant, severe burns, damages liver, kidneys, and lungs
3,3'-Dichlorobenzidine	DCB,	91-94-1	Used in manufacture of pigments for ink, textiles, plastics, and crayons.	Animal carcinogen, skin reactions.
2,4-Dichlorophenol	DCP	120-83-2	Feedstock for manufacture of 1,4-D and its derivatives (germicides, soil sterilants), used in mothproofing, antiseptics and seed disinfectants.	Suspected human carcinogen.
Diethylphthalate	DEP	84-66-2	Used as a vehicle in pesticidal sprays, solvent in perfumery.	Irritation to nose, throat, and chest. May cause nausea.
2,4-Dimethylphenol	2,4-Xylenol, 2,4-DMP	105-67-9	Used in manufacture of antioxidants, disinfectants, solvents, pharmaceuticals, insecticides, fungicides, plasticizers, rubber chemicals, and dyestuffs, and is an additive in lubricants, and gasolines. Also a component of coal tar.	Skin and eye irritant. May cause dizziness, nausea, vomiting, and exhaustion.
Dimethylphthalate	Phthalic acid, DMP, dimethyl ester	131-11-3	Used as plasticizer and insect repellant.	Irritation of nose and throat with coughing.
4,6-Dinitro-2-methylphenol	Dinitor-o-cresol, DNOC, 3,5-dinitro-o-cresol, 2-methyl-4,6-dinitrophenol	534-52-1	Used as herbicides and pesticide.	Fever, sweating headache, confusion. Elevated pulse, blood pressure. Orally lethal to humans.
2,4-Dinitrophenol	2,4-DNP	51-28-5	Used in manufacture of wood preservatives, pesticides, herbicides, explosives, and photographic developers.	Dizziness to death.
2,4-Dinitrotoluene	Dintrotoluol, 2,4-DNT	121-14-2	Used in preparation of polyurethane foams and plastics, used in manufacture of various explosives.	Animal carcinogen, affects nervous system and blood, causes coma.

Table 2.2 Semi-Volatile Organic Compounds (SVOCs) (Continued)

Compound	Synonyms	CAS No.	Use or Source	Toxicology (Why you're looking for it)
2,6-Dinitrotoluene	2,6-DNT	606-20-2	Used in preparation of polyurethane foams and plastics, used in manufacture of various explosives.	Animal carcinogen, affects nervous system and blood, causes coma.
Di-n-butylphthalate		84-74-2	Makes plastics flexible, found in insect repellents, glues, carpet backing, hair spray, nail polish, and rocket fuel.	No known harmful affects in humans. Causes cancer in rats.
Di-n-octylphthalate	DOP, benzenedicarboxylic acid.	117-84-0	Used as a plasticizer in manufacture of plastics.	Skin, eye, nose irritant. May cause death.
Fluoranthene	Benzo(k)fluorene, 1,2-benzacenaphthene, idryl.	206-44-0	Produced from heating coal and petroleum at high temperatures. Also a product of plant biosynthesis.	Toxic to freshwater and marine organisms. Not proven to be a carcinogen, yet is a PAH.
Fluorene	diphenylenemethane,	86-73-7	A PAH caused by incomplete combustion of coal, meat, oil, gas, garbage. Used to make dyes, plastics, and pesticides.	Attacks organs, not a carcinogen.
Hexachlorobenzene	Perchlorobenzene, HCB	118-74-1	Used as a fungicide, an additive for explosives, used in dye, and wood preservative manufacture.	Animal carcinogen, eye, nose irritant, causes nausea, vomiting, convulsions, and may cause coma.
Hexachloro-butadiene	Perchorobutadiene, 1,3-hexachlorobutadiene	87-68-3	Solvent for elastomers, a heat-transfer fluid, a transformer and hydraulic fluid.	Carcinogen (suspected human), may affect fetus, kidney and liver damage, burns skin and eyes.
Hexachlorocyclo-pentadiene	C-56®, HCCD, hex.	77-47-4	Flame retardent, used in manufacture of pesticides (aldrin, dieldrin, and endosulfan).	May cause death if swallowed or inhaled, skin and eye irritant, affects various organs.
Hexachloroethane	Perchloroethane, carbon hexachloride, HCE	67-72-1	Added to feed of ruminants, used in metal and alloy production.	Animal carcinogen, liver and skin irritant.
Indeno(1,2,3-c,d)pyrene	O-phenylpyrene, IP, 2,3-o-phenylenpyrene	193-39-5	Found in engine exhausts, cigarette smoke condensate, soot, coal tar pitch.	Animal carcinogen.

Table 2.2 Semi-Volatile Organic Compounds (SVOCs) (Continued)

Compound	Synonyms	CAS No.	Use or Source	Toxicology (Why you're looking for it)
Isophorone	3,3,5-Trimethyl-2-cyclohexene-1-one, trimethyl-cyclohexenone	78-59-1	Produced from acetone, it is used as solvent for finishes, lacquers, resins, pesticides, herbicides, fats, oils, and gums.	Carcinogen (animal suspected), headaches, nausea, can burn eyes, ENT irritant.
2-Methyl-naphthalene		91-57-6	Used in manufacture of dyes, resins, and Vitamin K.	Headache, nausea, and vomiting.
2-Methylphenol	ortho-cresol, o-cresol	95-48-7	Cresol is used for treating wood, insecticides, disinfectants. Found in wood and tobacco smoke, various foods including tomatos, butter, cheese, and bacon.	Irritates and burns skin, eyes, mouth, and throat. Abdominal pain, vomiting; coma; and death.
3-Methylphenol	meta-cresol, m-cresol	108-39-4		
4-Methylphenol	para-cresol, p-cresol	106-44-5		
Naphthalene	Moth balls, naphthalin, tar camphor.	91-20-3	Used as feedstock in manufacture of dyes. Used in manufacture of synthetic resins, smokeless powder, lampblack. Used as moth repellant.	Headache, nausea, sweating, and vomiting.
2-Nitroaniline		88-74-4	Used in manufacture of dyes, antioxidants, pharmaceuticals, and pesticides.	Affects blood's ability to carry oxygen, possible carcinogen, headaches, dizziness, death.
3-Nitroanaline		99-09-2		
4-Nitroaniline	Azoic diazo component 37, p-nitroaniline	100-01-6		
Nitrobenzene	Nitrobenzol, oil of mirbane, oil of bitter almonds	98-95-3	Used in manufacture of explosives and dyes. Used in shoe and floor polishes, leather dressings, and paint solvents. Used to mask unpleasant odors.	Blue tint to skin, rapid heart rate, low blood pressure, headache, lethargy, weakness, dizziness, and coma.
2-Nitrophenol	Hydroxynitobenzenes	88-75-5	Used in manufacture of dyestuffs and pesticides.	Affects blood's ability to carry oxygen, possible carcinogen, headaches, dizziness, death.
4-Nitrophenol	Hydroxynitobenzenes	100-02-7	Used in manufacture of dyestuffs and pesticides.	Affects blood's ability to carry oxygen, possible carcinogen, headaches, dizziness, death.
N-Nitroso-di-n-propylamine	N-nitroso-dipropylamine, N,N-dipropyl-nitrosamine.	621-64-7	Inadvertently produced during manufacturing process of some weed killers and rubber products.	Animal carcinogen, suspected human carcinogen.

Table 2.2 Semi-Volatile Organic Compounds (SVOCs) (Continued)

Compound	Synonyms	CAS No.	Use or Source	Toxicology (Why you're looking for it)
N-nitrosodi-phenylamine	N,N-diphenyl-nitrosamine, N-nitroso-N-phenyl-aniline, diphenylnitrosamine.	86-30-6	A mand made chemical no longer in production that was used in making rubber products (tires).	Animal carcinogen, suspected human carcinogen.
Pentachlorophenol	Penta, PCP, penchlorol	87-86-5	Bactericide, fungicide and slimicide used in wood preservatives, herbicide, molluscicide.	Irritant of nose, throat, and lungs. Can cause sneezing.
Phenanthrene		85-01-8	A PAH caused by incomplete combustion of coal, meat, oil, gas, garbage. Used to make dyes, plastics, and pesticides.	Attacks organs, not a carcinogen.
Phenol	Carbolic acid, phenic acid, phenyl hydrate, hydroxybenzene.	108-95-2	Used in manufacture of explosives, fertilizer, coke, illuminating gas, paints, paint removers, rubber, asbestos goods, wood preservatives. Used as a disinfectant.	Rapid heart rate, low blood pressure, cardiac failure, severe skin burns.
Pyrene	Benzo(d,e,f)phenanthre ne	129-00-0	Used in biochemical research.	Skin irritant, exhibits teratogenic effects.
Pyridine	Azine, azabenzene	110-86-1	Used as solvent in chemical industry. Used in manufacture of paints, explosives, and vitamins.	Nose and throat irritant, headache, dizziness, nausea, and vomitting.
1,2,4-Trichlorobenzene	Unsym-trichlorobenzene	120-82-1	Used as a dye carrier, heat transfer medium, degreaser, and lubricant.	Nose and throat irritant, tremors, increased heart rate.
2,4,5-Trichlorophenol	Dowcide 2	95-95-4	Used primarily as fungicide in pulp and paper industry.	Severe skin and eye irritant.
2,4,6-Trichlorophenol	Dowcide 2S, phenachlor	88-06-2	Fungicide, insecticide, herbicide, glue preservative.	Severe skin and eye irritant.

Table 2.3. Base/Neutral Compounds

Compound	Synonyms	CAS No.	Use or Source	Toxicology (Why you're looking for it)
Acenaphthene	1,8-ethylenenaphthalene	83-32-9	Occurs in coal tar (produced by coking of coal) and used in manufacture of some plastics, as insecticide and fungicide.	Skin irritant and may cause vomitting.
Acenaphthylene	Cyclopenta(de)-naphthalene	208-96-8	Coal tar processing.	Causes irritation to digestive and respitory system, skin, eyes.
Anthracene	Paranaphthalene	120-12-7	Used in insecticides and wood preservatives.	May cause mutations. Throat and nose irritant.
Benzo(a)anthracene	Benzphenanthrene, tetraphene, benzanthrene, BA	56-55-3	Found in exhaust, soot, coal and gas works emissions, and cigarette smoke. Also found in waxes, creosote, coal tar, and petroleum asphalt. Found in charcoal-broiled, barbecued, or smoked meats and fish and roasted coffee.	Animal carcinogen.
Benzo(a)pyrene	B(a)P (BAP), 3,4-benzpyrene	50-32-8	By-product of coal combustion. Also found in asphalt and cigarette smoke.	Animal carcinogen. Affects gastrointestinal system.
Benzo(b)fluoranthene	B (b) F, benz(e)acephenanthrylene	205-99-2	Formed during incomplete combustion of organic matter (fossil fuel, garbage). Component of creosote, coal tar pitch, and bituments.	Animal carcinogen.
Benzo(g,h,i)perylene		191-24-2	A PAH caused by incomplete combustion of coal, meat, oil, gas, garbage. Used in mfg. of explosives, plastics, pesticides, drugs. Used to make steroids, cholesterol, and bile acids.	No known harmful affects in humans.
Benzo(k)fluoranthene	Dibenzo(b,jk)fluorene	207-08-9	By-product of coal combustion. Also found in asphalt and cigarette smoke.	Affects gastrointestinal system. Animal carcinogen.
bis(2-Chloroethoxy) methane	BCEXM, dichlorethyl formal	111-91-1	Used in treatment of textiles, manufacture of polymers, insecticides, and as degreasing agents.	Toxicological properties have not been evaluated fully.
bis(2-Chloroethyl) ether	Dichloroethyl ether, dichloroether, BCEE, dichlorethyl oxide.	111-44-4	Used in manufacture of paint, varnish, lacquer, soap, and finish remover. Solvent for fats, oils, naphthalenes, greases, pectin, tar, and gum.	Animal carcinogen, coughing, tearing, retching, burns to skin and eyes, death.

Table 2.3. Base/Neutral Compounds (Continued)

Compound	Synonyms	CAS No.	Use or Source	Toxicology (Why you're looking for it)
bis(2-Ethylhexyl) phthalate	Di (2-ethylhexyl) phthalate	117-81-7	Used as a plasticizer for resins.	Animal carcinogen.
4-Bromophenyl-phenyl ether		101-55-3	Has been found in raw water, dringing water, and river water.	Skin contact leads to dermatitis.
Butylbenzylphthalate		85-68-7	Used as a plasticizer for polyvinyl resins.	Skin irritant, carcinogen.
Carbazole	9-Azafluorene, dibenzopyrrole	86-74-8	Used in dyes, explosives.	Irritant to skin, eyes, digestive system.
4-Chloroaniline	para-chloroaniline	106-47-8	Used to make dyes, other chemicals, insecticides and various industrial products.	Headaches, trouble breathing, weakness, bluish color of nose and lips, collapse, and death.
2-Chloronaphthalene		91-58-7	Wide use in industry - including in production of electric condensers, insulation of electric cables and wires, additives to pressure lubricants.	Causes chloracne (pustules, papules, cysts).
4-Chlorophenyl-phenyl ether	Benzene, 1-chloro-4-phenoxy	7005-72-3	Used as a dielectric fluid.	May damage digestive track. May cause irritation to eyes and skin.
2,2'-oxybis (1-chloropropane)	bis(2-Chloroisopropyl) ether bis(2-Chloro-1-methylethyl) ether	108-60-1	Used in manufacture of dyes, resins, pharmaceuticals, and textiles. Also used at fungicide and insecticide in wood preservatives. Used in manufacture of glycols.	Inhalation, ingestion, or skin contact may cause severe injury or death.
Chrysene	1,2-Benzophenanthrene	218-01-9	Found in gasoline and diesel exhaust, cigarette smoke, and coal tar.	Animal carcinogen, skin rashes, pigment change.
Dibenz(a,h)anthracene	DBA, 1,2,3,6-dibenzanthracene	53-70-3	Formed during incomplete combustion of organic matter (fossil fuel, garbage). Component of creosote, coal tar pitch, and bitumens.	Animal carcinogen.
Dibenzofuran	2,2'-Biphenylen oxide	132-64-9	Used as an insecticide.	Toxic but details not known.

Table 2.3. Base/Neutral Compounds (Continued)

Compound	Synonyms	CAS No.	Use or Source	Toxicology (Why you're looking for it)
1,2-Dichlorobenzene	1,2-DCB	95-50-1	Used as solvent and in manufacture of dyes, herbicides, and degreasers.	Blood cell damage, skin and eye irritant, severe burns, damages liver, kidneys, and lungs.
1,3-Dichlorobenzene	1,3-DCB	541-73-1	Contaminant of 1,2 and 1,4-DCB formulation	Blood cell damage, skin and eye irritant, severe burns, damages liver, kidneys, and lungs.
1,4-Dichlorobenzene	1,4-DCB	106-46-7	Air deodorant and insecticide.	Blood cell damage, skin and eye irritant, severe burns, damages liver, kidneys, and lungs.
3,3'-Dichlorobenzidine	DCB,	91-94-1	Used in manufacture of pigments for ink, textiles, plastics, and crayons.	Animal carcinogen, skin reactions.
Diethylphthalate	DEP	84-66-2	Used as a vehicle in pesticidal sprays, solvent in perfumery.	Irritation to nose, throat, and chest. May cause nausea.
Dimethylphthalate	Phthalic acid, DMP, dimethyl ester	131-11-3	Used as plasticizer and insect repellant.	Irritation of nose and throat with coughing.
2,4-Dinitrotoluene	Dintrotoluol, 2,4-DNT	121-14-2	Used in preparation of polyurethane foams and plastics, used in manufacture of various explosives.	Animal carcinogen, affects nervous system and blood, causes coma.
2,6-Dinitrotoluene	2,6-DNT	606-20-2	Used in preparation of polyurethane foams and plastics, used in manufacture of various explosives.	Animal carcinogen, affects nervous system and blood, causes coma.
Di-n-butylphthalate		84-74-2	Makes plastics flexible, found in insect repellents, glues, carpet backing, hair spray, nail polish, and rocket fuel.	No known harmful affects in humans. Caused cancer in rats.
Di-n-octylphthalate	DOP, benzenedicarboxylic acid.	117-84-0	Used as a plasticizer in manufacture of plastics.	Skin, eye, nose irritant. May cause death.
Fluoranthene	Benzo(k)fluorene, 1,2-benzacenaphthene, idryl.	206-44-0	Produced from heating coal and petroleum at high temperatures. Also a product of plant biosynthesis.	Toxic to freshwater and marine organisms. Not proven to be a carcinogen, yet is a PAH.

Table 2.3. Base/Neutral Compounds (Continued)

Compound	Synonyms	CAS No.	Use or Source	Toxicology (Why you're looking for it)
Fluorene	diphenylenemethane,	86-73-7	A PAH caused by incomplete combustion of coal, meat, oil, gas, garbage. Used to make dyes, plastics, and pesticides.	Attacks organs, not a carcinogen.
Hexachlorobenzene	Perchlorobenzene, HCB	118-74-1	Used as a fungicide, an additive for explosives, used in dye, and wood preservative manufacture.	Animal carcinogen, eye, nose irritant, causes nausea, votmiting, convulsions, and may cause coma.
Hexachlorobutadiene	Perchorobutadiene, 1,3-hexachlorobutadiene	87-68-3	Solvent for elastomers, a heat-transfer fluid, a transformer and hydraulic fluid.	Carcinogen (suspected human), may affect fetus, kidney and liver damage, burns skin and eyes.
Hexachlorocyclo-pentadiene	C-56®, HCCD, hex.	77-47-4	Flame retardent, used in manufacture of pesticides (aldrin, dieldrin, and endosulfan)	May cause death if swallowed or inhaled, skin and eye irritant, affects various organs.
Hexachloroethane	Perchloroethane, carbon hexachloride, HCE	67-72-1	Added to feed of ruminants, used in metal and alloy production.	Animal carcinogen, liver and skin irritant.
Indeno(1,2,3-c,d)pyrene	O-phenylpyrene, IP, 2,3-o-phenylenpyrene	193-39-5	Found in engine exhausts, cigarette smoke condensate, soot, coal tar pitch.	Animal carcinogen.
Isophorone	3,3,5-Trimethyl-2-cyclohexene-1-one, trimethylcyclohexenone.	78-59-1	Produces from acetone it is used as solvent for finishes, lacquers, resins, pesticides, herbicides, fats, oils, and gums.	Carcinogen (animal suspected), headaches, nausea, can burn eyes, ENT irritant.
2-Methylnaphthalene		91-57-6	Used in manufacture of dyes, resins, and Vitamin K.	Headache, nausea, and vomiting.
Naphthalene	Moth balls, naphthalin, tar camphor.	91-20-3	Used as feedstock in manufacture of dyes. Used in manufacture of synthetic resins, smokeless powder, lampblack. Used as moth repellant.	Headache, nausea, sweathing, and vomiting.

Table 2.3. Base/Neutral Compounds (Continued)

Compound	Synonyms	CAS No.	Use or Source	Toxicology (Why you're looking for it)
2-Nitroaniline		88-74-4	Used in manufacture of dyes, antioxidants, pharmaceuticals, and pesticides.	Affects blood's ability to carry oxygen, possible carcinogen, headaches, dizziness, death.
3-Nitroanaline		99-09-2		
4-Nitroaniline	Azoic diazo component 37, p-nitroaniline.	100-01-6		
Nitrobenzene	Nitrobenzol, oil of mirbane, oil of bitter almonds	98-95-3	Used in manufacture of explosives and dyes. Used in shoe and floor polishes, leather dressings, and paint solvents. Used to mask unpleasant odors.	Blue tint to skin, rapid heart rate, low blood pressure, headache, lethargy, weakness, dizziness, and coma.
N-Nitroso-di-n-propylamine	N-nitrosodipropylamine, N,N-dipropylnitrosamine.	621-64-7	Inadvertently produced during manufacturing process of some weed killers and rubber products.	Animal carcinogen, human suspected.
N-nitrosodiphenylamine	N,N-diphenylnitrosamine, N-nitroso-N-phenylaniline, diphenylnitrosamine.	86-30-6	A manmade chemical no longer in production that was used in making rubber products (tires).	Animal carcinogen, human suspected.
Phenanthrene		85-01-8	A PAH caused by incomplete combustion of coal, meat, oil, gas, garbage. Used to make dyes, plastics, and pesticides.	Attacks organs, not a carcinogen.
Pyrene	Benzo(d,e,f)phenanthrene	129-00-0	Used in biochemical research.	Skin irritant, exhibits teratogenic effects.
Pyridine	Azine, azabenzene	110-86-1	Used as solvent in chemical industry. Used in manufacture of paints, explosives, and vitamins.	Nose and throat irritant, headache, dizziness, nausea, and vomiting.
1,2,4-Trichlorobenzene	Unsym-trichlorobenzene	120-82-1	Used as a dye carrier, heat transfer medium, degreaser, and lubricant.	Nose and throat irritant, tremors, increased heart rate.

Table 2.4. Acid Compounds

Compound	Synonyms	CAS No.	Use or Source	Toxicology (Why you're looking for it)
4-Chloro-3-methylphenol	4-chloro-m-cresol	59-50-7	Germicide, preservative for glues, gums, paints, inks, textile & leather goods.	Can harm eyes and skin; toxic to fish.
2-Chlorophenol		95-57-8	Used in manufacture of fungicides, pesticides, herbicides, disinfectants, wood and glue preservatives.	Severe irritation and burns to nose and throat; headache, dizziness, lung damage.
2,4-Dichlorophenol	DCP	120-83-2	Feedstock for manufacture of 1,4-D and its derivatives (germicides, soil sterilants), used in mothproofing, antiseptics, and seed disinfectants.	Suspected human carcinogen.
2,4-Dimethylphenol	2,4-Xylenol, 2,4-DMP	105-67-9	Used in manufacture of antioxidants, disinfectants, solvents, pharmaceuticals, insecticides, fungicides, plasticizers, rubber chemicals, and dyestuffs, and is an additive in lubricants, and gasolines. Also a component of coal tar.	Skin and eye irritant. May cause dizziness, nausea, vomiting,and exhaustion.
Dimethyl-phthalate	Phthalic acid, DMP, dimethyl ester	131-11-3	Used as plasticizer and insect repellant.	Irritation of nose and throat with coughing.
4,6-Dinitro-2-methylphenol	Dinitor-o-cresol, DNOC, 3,5-dinitro-o-cresol, 2-methyl-4,6-dinitrophenol	534-52-1	Used as herbicide and pesticide.	Fever, sweathing headache, confusion. Elevated pulse, blood pressure. Orally lethal to humans.
2,4-Dinitrophenol	2,4-DNP	51-28-5	Used in manufacture of wood preservatives, pesticides, herbicides, explosives, and photographic developers.	Dizziness to death.
2-Methylphenol	ortho-cresol, o-cresol	95-48-7		Irritates and burns skin, eyes, mouth, and throat. Abdominal pain, vomiting, coma, and death.
3-Methylphenol	meta-cresol, m-cresol	108-39-4		
4-Methylphenol	para-cresol, p-cresol	106-44-5		
2-Nitrophenol	Hydroxynitobenzenes	88-75-5	Used in manufacture of dyestuffs and pesticides.	Affects blood's ability to carry oxygen, possible carcinogen, headaches, dizziness, death.

Table 2.4. Acid Compounds (Continued)

Compound	Synonyms	CAS No.	Use or Source	Toxicology (Why you're looking for it)
4-Nitrophenol	Hydroxynitobenzenes	100-02-7	Used in manufacture of dyestuffs and pesticides.	Affects blood's ability to carry oxygen, possible carcinogen, headaches, dizziness, death.
Pentachloro-phenol	Penta, PCP, penchlorol	87-86-5	Bactericide, fungidice and slimicide used in wood preservatives, herbicide, molluscicide.	Irritant of nose, throat, and lungs. Can cause sneezing.
Phenol	Carbolic acid, phenic acid, phenyl hydrate, hydroxybenzene.	108-95-2	Used in manufacture of explosives, fertilizer, coke, illuminating gas, paints, paint removers, rubber, asbestos goods, wood preservatives. Used as a disinfectant.	Rapid heart rate, low blood pressure, cardiac failure, severe skin burns.
2,4,5-Trichlorophenol	Dowcide 2	95-95-4	Used primarily as a fungicide in pulp and paper industry.	Severe skin and eye irritant.
2,4,6-Trichlorophenol	Dowcide 2S, phenachlor	88-06-2	Fungicide, insecticide, herbicide, glue preservative.	Severe skin and eye irritant.

SVOCs can be divided into another category, most often called polynuclear aromatic hydrocarbons (PAHs), but sometimes referred to as PNAs. I have also seen these compounds referred to as polycyclic, as opposed to polynuclear. Regardless of their name, PAHs are abundant in the environment. They are a product of incomplete combustion of organic compounds including coal, oil, gas, diesel, wood, volcanoes, garbage, tobacco, and meat. PAHs are the concern that people have when meat is over-cooked. They are abundant in roofing tar, coal tar, asphalt, crude oil, and creosote and are used to make dyes, plastics, and pesticides.

While there are hundreds of PAHs, there are 16 that are of primary concern to the environment (see Table 2.5). These 16 are a focus for four reasons. First, there is more toxicological information available for them, as several are carcinogens. Second, it is suspected that they are more harmful and they exhibit harmful effects that are representative of PAHs. Third, there is an increased chance of exposure to these 16 PAHs. And finally, these PAHs are more abundant at hazardous waste (NPL) sites. Once again, keep in mind that PAHs are BNAs and SVOCs.

Total Petroleum Hydrocarbons

To discuss total petroleum hydrocarbons (TPH), let's start with crude oil. Crude oil is the root for many chemicals for which we sample. Crude oil is often pictured as a thick, black, oozing material, but it can also be quite fluid. Crude oil can produce not only motor oil, but gasoline, jet fuel, kerosene, diesel, mineral spirits, and tar. There are a variety of other chemicals that come from crude oil as well, and they are all separated through the distilling process. (See Figure 2.2.)

In high school chemistry, there was an experiment conducted where we distilled water. We boiled water in a flask with a stopper in it. A glass tube ran from a hole in the stopper to another flask. As the water boiled, the vapor or steam traveled through the glass tube toward the other flask. As it cooled, it returned to its liquid phase (water) and poured into the flask. The water distilling process occurs at 212°F, the boiling point of water. The many chemicals that comprise crude oil boil at different temperatures. In a refinery, the crude oil is boiled at varying degrees and the varying chemicals "pour out" into a "flask." The many products of crude oil boil or distill at different temperatures based on their chemical composition, meaning the number of carbon atoms in their molecular structure.

TPH is a good screening method to assess the presence of petroleum. There are three portions of hydrocarbons for which you would screen: gasoline range organics (GRO), diesel range organics (DRO), and oil range organics (ORO).

Table 2.5. Polynuclear Aromatic Hydrocarbons (PAHs)

Compound	Synonyms	CAS No.	Toxicology (Why you're looking for it)
Acenaphthene	1,8-ethylenenaphthalene	83-32-9	Skin irritant and may cause vomitting.
Acenaphthylene	Cyclopenta(de)naphthalene	208-96-8	Causes irritation to digestive and respitory system, skin, eyes.
Anthracene	Paranaphthalene	120-12-7	May cause mutations. Throat and nose irritant.
Benzo(a)anthracene	Benzphenanthrene, tetraphene, benzanthrene, BA	56-55-3	Animal carcinogen.
Benzo(a)pyrene	B(a)P (BAP), 3,4-benzpyrene	50-32-8	Animal carcinogen. Affects gastrointestinal system.
Benzo(b)fluorene	B (b) F, benz(e)acephenanthrylene	205-99-2	Animal carcinogen.
Benzo(g,h,i)perylene		191-24-2	No known harmful affects in humans.
Benzo(k)fluorene	Dibenzo(b,jk)fluorene	207-08-9	Affects gastrointestinal system. Animal carcinogen.
Chrysene	1,2-Benzophenanthrene	218-01-9	Animal carcinogen, skin rashes, pigment change.
Dibenzo(a,h)anthracene	DBA, 1,2,3,6-dibenzanthracene	53-70-3	Animal carcinogen.
Fluoranthene	Benzo(k)fluorene, 1,2-benzacenaphthene, idryl.	206-44-0	Toxic to freshwater and marine organisms. Not proven to be a carcinogen yet is a PAH.
Fluorene	diphenylenemethane,	86-73-7	Attacks organs, not a carcinogen.
Indeno(1,2,3-cd)pyrene	0-phenylpyrene, IP, 2,3-o-phenylenpyrene	193-39-5	Animal carcinogen.
Phenanthrene		85-01-8	Attacks organs, not a carcinogen.
Pyrene	Benzo(d,e,f)phenanthrene	129-00-0	Skin irritant, exhibits teratogenic effects.
Naphthalene	Moth balls, naphthalin, tar camphor.	91-20-3	Headache, nausea, sweathing, and vomiting.

Figure 2.2. Hydrocarbon Range

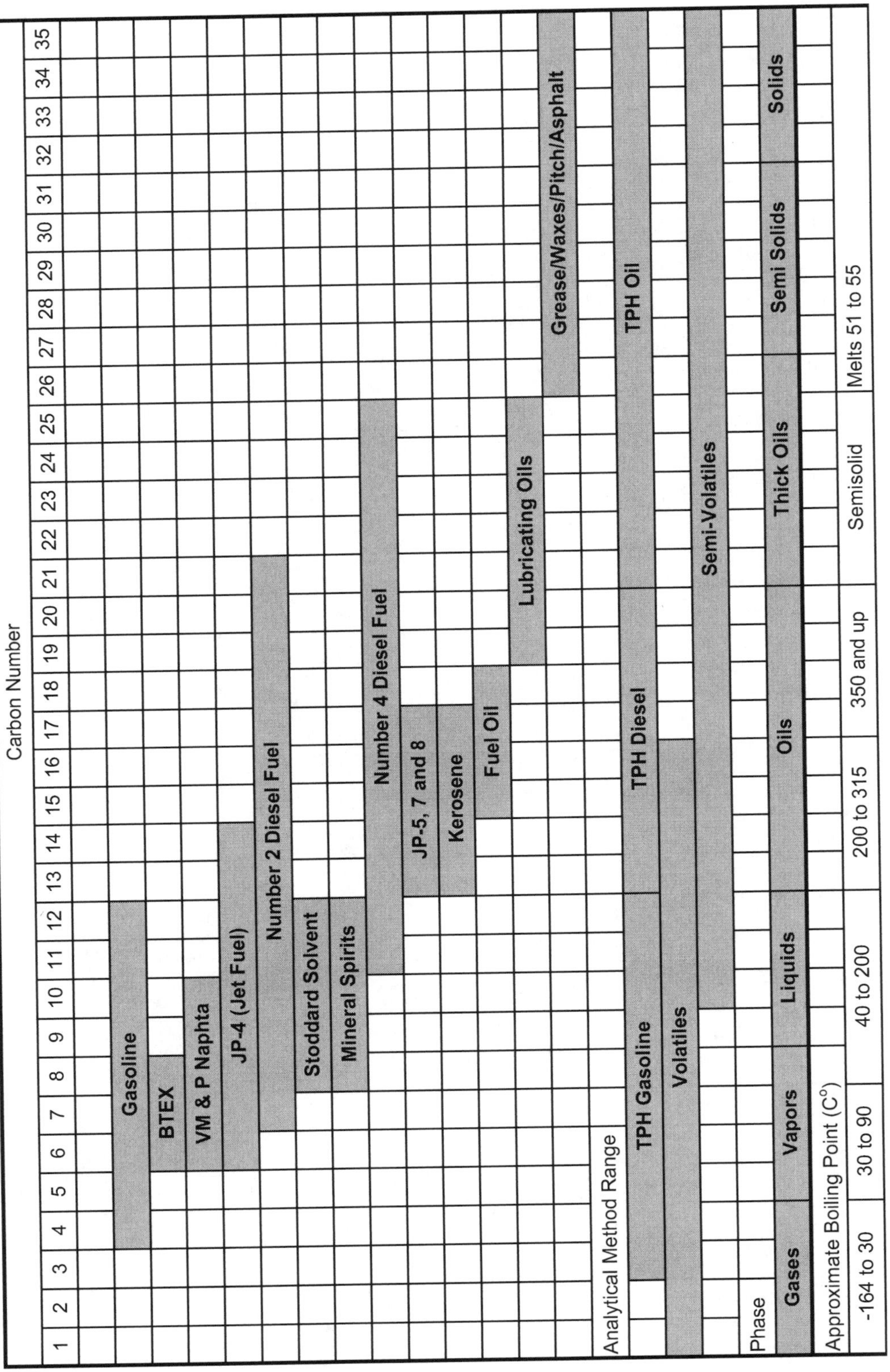

	Carbon Number
Gasoline	4–12
BTEX	6–8
VM & P Naphta	6–10
JP-4 (Jet Fuel)	6–14
Number 2 Diesel Fuel	7–21
Stoddard Solvent	8–12
Mineral Spirits	8–12
Number 4 Diesel Fuel	11–25
JP-5, 7 and 8	13–17
Kerosene	13–17
Fuel Oil	15–18
Lubricating Oils	19–25
Grease/Waxes/Pitch/Asphalt	26–35
Analytical Method Range	
TPH Gasoline	3–12
TPH Diesel	13–20
TPH Oil	21–35
Volatiles	1–16
Semi-Volatiles	10–35
Phase	
Gases	1–4
Vapors	5–8
Liquids	9–12
Oils	13–20
Thick Oils	21–26
Semi Solids	27–31
Solids	32–35
Approximate Boiling Point (C°)	
-164 to 30	1–4
30 to 90	5–7
40 to 200	8–12
200 to 315	13–16
350 and up	17–20
Semisolid	21–25
Melts 51 to 55	26–35

Courtesy of First Environmental Laboratories, Inc. Naperville, IL

GRO analysis provides the total compounds, measured in mg/kg (parts per million). DRO, naturally, provides the total compounds in the diesel range. This range can be extended to include hydrocarbons containing 40 carbons. The ORO is less frequently screened for because those components may be included in the DRO analysis. When you analyze for GRO and DRO, you can't sum the results because GRO includes some lighter compounds in the diesel range just as DRO results would include some heavier GRO compounds.

TPH with GRO and DRO is SW-846 Method 8015. There are other methods that have been used but are becoming obsolete due to the use of freon in the extraction process. Methods 413.1, 413.2, and 418 use freon to extract the compounds from the sample. Freon is an air contaminant and continued use of the gas is restricted. Extraction is necessary to get contaminants out of the matrix they are in. If naphthalene, a semi-volatile in diesel, is in soil, you have to extract it from the soil in order to quantify it. Laboratories have to extract naphthalene using chemicals to which naphthalene is "attracted." That has been freon in the past. Another commonly used extractor is methylene chloride.

PCBs, Pesticides, and Herbicides

Polychlorinated biphenyls (PCBs) are an often sought after compound when sampling. They were used widely from the 1930s through to the early 1970s. Their purpose was to act as an insulator in oils so the oils could be used at higher temperatures and/or pressures without catching fire. Electrical transformers often have PCBs in their oil (dielectric fluid), so when sampling at a site with transformers, attention should be given to the condition of the transformer and indication of stains beneath or around it. PCBs were also used in other lubricating and hydraulic oils, and may have been used in blue ink.

When PCB analytical results are presented, they are typically presented in a list of Aroclors®. The most common Aroclors are 1016, 1242, 1248, 1254, 1260, 1262, and 1268. In these numbers, the first two digits represent the number of carbon atoms in the PCB, and the last two represent the percentage of chlorine. So, Aroclor 1242 has 12 carbon atoms and is 42% chlorine.

Sampling for pesticides and herbicides primarily is uncommon because they are not as prevalent in the environment as VOCs, SVOCs, or metals. In fact, when disposing of waste, you can often sign a form indicating that pesticides and herbicides were never used at the source property therefore, analysis for these compounds is not necessary. However, at all Superfund sites at which I have sampled, I have had to collect samples for pesticides and herbicides. There is nothing unique in sampling for these compounds, and they usually only require an additional sample container or two to provide sufficient quantity for analysis. Table 2.6 presents the standard suite of pesticides and herbicides, and some common products in which you may find these chemicals.

Table 2.6. Pesticides and Herbicides

Compound	Synonyms	CAS No.	Use or Source	Toxicology (Why you're looking for it)
Pesticides/PCBs				
Aldrin		309-00-2	Insecticide on cotton, for termites. Banned in 1987.	Probable carcinogen, also causes dizziness, convulsions, affects nervous system.
alpha-BHC	Hexachlorocyclohexanes	319-85-6	Used as insecticide on fruits and vegetables. (US production of Lindane banned in 1977.) Still imported.	Affects liver and kidney.
beta-BHC	Hexachlorocyclohexanes	319-85-7		Affects liver and kidney.
gamma-BHC	Hexachlorocyclohexanes	319-85-8		Affects liver and kidney.
delta-BHC	Hexachlorocyclohexanes, Lindane	58-89-9		Affects liver and kidney.
alpha-Chlordane	Chlorindan, Chlor Kil, Dowchlor, Octachlor and Velsicol 1068.	5103-71-9	Pesticide for crops (corn and citrus), home lawn and garden. Banned in 1983 except as termiticide. Banned as termiticide in 1988.	Affects liver and nervous system.
gamma-Chlordane		5566-34-7		Affects liver and nervous system.
4,4'-DDD	DDD	72-54-8	Breakdown product of DDT	More of a threat to animals than to humans. Probable carcinogen.
4,4'-DDE	DDE	72-55-9	Breakdown product of DDT	More of a threat to animals than to humans. Probable carcinogen.
4,4'-DDT	DDT, dichlorodiphenyltrichloroethane	50-29-3	Banned insecticide; used to kill mosquitos.	More of a threat to animals than to humans. Probable carcinogen.
Dieldrin		60-57-1	Breakdown of Aldrin.	Probable carcinogen, also causes dizziness, convulsions, affects nervous system.
Endosulfan I	Cyclodan, Endosol, Insectophene, Malix, Thiodan, Thiosulfan.	115-29-7	Insecticide for food crops and wood.	Dizziness to death.
Endosulfan II		33213-65-9	Insecticide for food crops and wood.	Dizziness to death.

Table 2.6. Pesticides and Herbicides (Continued)

Compound	Synonyms	CAS No.	Use or Source	Toxicology (Why you're looking for it)
Endosulfan Sulfate		1031-07-8	Breakdown of Endosulfan I	Dizziness to death.
Endrin	mendrin	72-20-8	Banned insecticide.	Affects liver and central nervous system.
Endrin Aldehyde		7421-93-4	Impurity of Endrin	Affects central nervous system.
Endrin Ketone		53494-70-5	Impurity of Endrin	Affects central nervous system.
Heptachlor	Termide®, Heptagran®, Drinox®.	76-44-8	Termiticide, banned in 1988	Affects liver.
Heptachlor Epoxide		1024-57-3	Breakdown of Heptachlor	Affects liver.
Lindane	benzene hexachloride.	58-89-9	Insecticide used on cattle, lumber, and gardens. Also used in bait for rodents.	Affects liver, kidney, central nervous system, and gastrointestinal system.
Mathoxychlor	DMDT, Marlate®, or Metox®.	72-43-5	Insecticide against flies, mosquitos, cockroaches, chiggers. Used on apples, potatoes, and tomatoes and other crops, livestock, grains, gardens, and pets.	Affects reproductive system by acting as a natural sex hormone.
Toxaphene	Alltox, chlorinated-camphene, phenacide, toxaikil, toxadust.	8001-35-2	Insecticide used on cotton and cattle.	Affects liver, kidney or thyroid gland.
Aroclor 1016	PCBs	12674-11-2	Used in oils as insulator to minimize fires. Found in transformers, fluorescent light balasts, hydraulic oil.	May cause liver damage and skin acne in high exposure. Animal carcinogen.
Aroclor 1221	PCBs	11104-28-2	Used in oils as insulator to minimize fires. Found in transformers, fluorescent light balasts, hydraulic oil.	May cause liver damage and skin acne in high exposure. Animal carcinogen.

Table 2.6. Pesticides and Herbicides (Continued)

Compound	Synonyms	CAS No.	Use or Source	Toxicology (Why you're looking for it)
Aroclor 1232	PCBs	11141-16-5	Used in oils as insulator to minimize fires. Found in transformers, fluorescent light balasts, hydraulic oil.	May cause liver damage and skin acne in high exposure. Animal carcinogen.
Aroclor 1242	PCBs	53469-21-9	Used in oils as insulator to minimize fires. Found in transformers, fluorescent light balasts, hydraulic oil.	May cause liver damage and skin acne in high exposure. Animal carcinogen.
Aroclor 1248	PCBs	12672-29-6	Used in oils as insulator to minimize fires. Found in transformers, fluorescent light balasts, hydraulic oil.	May cause liver damage and skin acne in high exposure. Animal carcinogen.
Aroclor 1254	PCBs	11097-69-1	Used in oils as insulator to minimize fires. Found in transformers, fluorescent light balasts, hydraulic oil.	May cause liver damage and skin acne in high exposure. Animal carcinogen.
Aroclor 1260	PCBs	111096-82-5	Used in oils as insulator to minimize fires. Found in transformers, fluorescent light balasts, hydraulic oil.	May cause liver damage and skin acne in high exposure. Animal carcinogen.

Table 2.6. Pesticides and Herbicides (Continued)

Compound	Synonyms	CAS No.	Use or Source	Toxicology (Why you're looking for it)
Herbicides				
Alachlor		15972-60-8	Herbicide for grasses & broadleaf weeds on row crops.	Affects eyes, liver, kidney, and spleen
Atrazine	Aatrex, Fenamine, Primex, Weedex A, Triazine A 1294.	1912-24-9	Used on row crops.	Affects cardiovascular and reproductive systems.
Carbofuran	Bay 70143, Furodan, Niagara 10242, Yaltox.	1563-66-2	Soil fumigant used on rice and alfalfa. Broad use insecticide and nematocide.	Affects liver.
2,4-D	2,4-dichlorophenoxyacetic acid	94-75-7	Used on row crops, broadweed grasses (Weed-B-Gone®).	Affects liver or kidney, acute poisoning can result in death.
Dalapon	2,2-Dichloropropanoic acid	75-99-0	Used on right-of-ways.	Minor kidney changes, affects CNS.
1,2-Dibromo-3-chloropropane	Fumagone, Fumazone, Nemagone, Nemafume,.	96-12-8	Soil fumigant used on soybeans, cotton, pineapples, and orchards.	Affects reproductive system
Dinoseb	Aatox, Dibutox, Kiloseb, Knoxweed, Laseb, Nitropone.	88-85-7	Used on soybeans and vegetables.	Affects reproductive system
Diquat	Aero Titanic, Preeglone, Reglone.	85-00-7 or 2764-72-9	General use.	Causes cataracts, dizziness, may be fatal if swallowed.
Endothall	Aquathol, Hydout, Hydrothol.	145-73-3	General use.	Affects gastrointestinal system.
Picloram	ATCP, Borolin, Chloramp, Tordon-10K.	1918-02-1	General use.	Affects liver.
Simazine	Herbex, Princep, Rodocon, Taphazine.	122-34-9	General use.	Affects blood.
2,4,5-TP (Silvex)	2,4,5-T, Reddon, Trioxon, Weedone.	93-72-1	Banned herbicide.	Affects liver.

Inorganic Chemistry

Inorganic chemistry is the study of the chemical compounds and elements that do not contain carbon. Some characteristics of inorganics are as follows:

1. They have high boiling and melting points,
2. They are inflammable,
3. They are soluble in water (salt water), and
4. They are insoluble in organic solvents (a degreaser takes away grease and not the metal).

The primary inorganics for which we sample are metals and cyanide. Sometimes chloride, nitrate, and sulfate are listed with inorganics, but in this book they will be discussed in the Wet Chemistry section.

Metals

Table 2.7 provides a list of metals and cyanide, their CAS numbers, and their possible toxicological effects. It is not often that you have to analyze for a full suite of metals. Typically, you will have to analyze for the "RCRA 8," those metals that are most prevalent and/or most harmful in the environment. The RCRA 8 metals are arsenic, barium, cadmium, chromium, lead, mercury, selenium, and silver. For disposal purposes, waste materials have to be analyzed for the RCRA 8 metals.

Of the eight RCRA metals, chromium often receives considerable media attention. There are several considerations when sampling for chromium. First, chromium is present in two forms; trivalent chromium and hexavalent chromium. Trivalent chromium or Chromium (III) is a nutrient and necessary for our bodies in small doses. Hexavalent chromium or chromium (VI) causes cancer and various other health problems. Chromium occurs naturally in crude oil and is produced from chromite ore. Sources of chromium (III) and (VI) include chrome plating, steel production, and pigmenting operations for paints, rubber, and plastics. Chromium (VI) is commonly used in lead chromate, which is the yellow pigment used for painting traffic lines.

When you sample for chromium (VI) in groundwater, it is important to remember its holding time is twenty-four hours. Chromium (VI) is readily soluble in water but chromium (III) is not. When chromium (VI) is exposed to acid or organic matter, it reduces to chromium (III). When you sample groundwater for metals, the sample container must have nitric acid in it to prevent chromium (VI) from reducing to (III) in that container. Also, the longer the groundwater is in the container, the longer organic matter has to reduce the (VI) to

Table 2.7. Metals

Compound	CAS No.	Use or Source	Toxicology (Why you're looking for it)
Aluminum	7429-90-5	Used in fireworks, foil, to produce glass, rubber, also used as a food additive.	High doses may cause respitory problems.
Antimony	**7440-36-0**	Used in lead batteries, solder, pewter, and bearings and as an alloy with other metals	Affects circulatory system
Arsenic	7440-38-2	Combined with sulfer, oxygen, and chlorine to form inorganic arsenic compounds (used for wood treatment). When combined with carbon and hydrogen, used for pesticides (moslty for cotton).	Affects circulatory and reproductive systems. Causes death in high doses. Inorganic compounds are more toxic than organic compounds.
Barium	7440-39-3	Used to make drilling mud, paint, bricks, tiles, glass, and rubber.	Affects heart, kidneys, stomach, liver, and other organs.
Beryllium	7440-41-7	Used to make nuclear weapons and reactors, aircraft, and mirrors. Alloys are used in vehicles and sports equipment.	Causes lung damage (scarring) when inhaled. Some people develop acute or chronic beryllium disease.
Boron	7440-42-8	Used to make glass, fire retardants, leather tanning industries, cosmetics, photographic materials, soaps and cleaners, and for high-energy fuel. Some pesticides used for cockroach control and some wood preservatives also contain borates.	Affects reproductive system. Irritant to eyes, nose, and throat.
Cadmium	7440-43-9	Used to make batteries, pigments, metal coatings, and plastics.	Severely damages lungs to point of death. Severely damages stomach. Can build up in kidneys.
Calcium	7440-70-2	Used in photoelectric cells, vacuum tubes.	May affect central nervous system, gastrointestinal system.

Table 2.7. Metals (Continued)

Compound	CAS No.	Use or Source	Toxicology (Why you're looking for it)
Chromium	7440-47-3	Used to make steel.	
Chromium $^{+3}$	16065-83-1	Used in plating, dyes, tanning and as wood preservative.	An essential nutrient to help process fats and sugars.
Chromium $^{+6}$	18540-29-9	Used in plating, dyes, tanning and as wood preservative.	Causes cancer. Affects nose and various organs.
Cobalt	7440-48-4	Used to make metals. Used as pigment, drier for paints and enamels.	As vitamin B12, an essential nutrient. Otherwise can affect lungs.
Copper	7440-50-8	Used in the penny, wiring, fancy roofs, water pipes. Compounds used as mildew treatment on plants, wood, leather and fabric preservative.	Causes irritation of the nose, mouth and eyes, vomiting, diarrhea, stomach cramps, and nausea. Also an essential nutrient.
Cyanide (amenable)	57-12-5	Cyanide (hydrogen cyanide) is used in electroplating, metallurgy, production of chemicals, photographic development, making plastics, fumigating ships, and some mining processes. Produced naturally in the environment.	From headaches to death. Affects thyroid gland, heart, brain, and blood.
Iron	15438-31-0	Used to make steel.	Chronic exposure causes lung, liver and gastrointestinal damge.
Lead	7439-92-1	Used to make batteries, amunition, x-ray shields, and metal products. Formally used in paints.	Can damage the nervous system, kidneys, and reproductive system.
Magnesium	7439-95-4	Used to make light alloys, optical mirrors, in pyrotechnics, for flash bulbs & flares.	Affects central nervous system and digestive tract.
Manganese	7439-96-5	Used to make steel and other alloys. Organic compounds used to make pesticides.	An essential nutrient in small does. Exposure can affects central nervous system.

Table 2.7. Metals (Continued)

Compound	CAS No.	Use or Source	Toxicology (Why you're looking for it)
Mercury	7439-97-6	Used in thermometers, barometers, batteries, old dental fillings. Used to make caustic soda and chlorine gas.	Affects central nervous system. May damage brain, kidneys, and developing fetus.
Nickel	7440-02-0	Used to make coins (e.g., nickels), stainless steel, other alloys, jewellery, plating.	May cause allergic reaction with skin (from jewellery), likely causes lung and sinus cancer.
Potassium	7440-09-7	Used in organic sytheses.	Affects liver, kidney, lungs, and skin (when wet).
Selenium	7782-49-2	Used in dandruff shampoos (selenium sulfide), used to clean guns (selenious acid).	An essential nutrient. Overexposure causes brittle hair, deformed nails, and numbness.
Silver	7440-22-4	Used for silverware, coins, jewellery, electronics, photo development, as an antibacterial agent and disinfectant.	Causes argyria (skin and organs turn blue-gray). Causes lung and throat irritations.
Sodium	7440-23-5	Used to manufacture various salts, sodium lamps, and photoelectric cells.	Causes skin and gastrointestinal burns, respitory tract irritation.
Thallium	7440-28-0	Used for electronics in the semiconductor industry, special glass, and certain medical procedures.	Affects nervous system, lungs, gastrointestinal system, heart, kidneys and causes hair loss.
Vanadium	7440-62-2	Used to make steel and alloys for auto and aircraft parts. Also used to make rubber, cermamics, and plastics.	Causes lung irritation, chest pain, and coughing.
Zinc	7440-66-6	Used to make brass and bronze, with copper to make pennies. Used make paint, rubber, dye, wood preservatives, and ointments (diaper rash ointment).	An essential nutrient. To little causes growth problems in adolescents and in-utero babies. Too much causes cramps, vomitting, and nausea.

(III). Analytical results for chromium will be presented as chromium, total; chromium (III); and chromium (VI). The sum of chromium (III) and chromium (VI) is chromium, total.

Cyanide can be presented in two ways: total and amenable. Total is the amount of cyanide that is actually in a sample. Amenable (or free) is the amount that can "be released" to harm human health or the environment. Typically you will see remediation objectives for amenable cyanide rather than total cyanide. When you sample groundwater, the container for cyanide is preserved with sodium hydroxide (NaOH), which is a base. When the sample is analyzed, the lab will acidify the groundwater to release the cyanide. Many analyses for other parameters follow this procedure of acidifying to release or extract contaminants. If you suspect that the material you submit has high cyanide concentrations, mark that on the chain-of-custody (COC) to alert the lab workers. They should run the cyanide sample first to assess that concentration. Also, it alerts them that any other acidifying they do should be conducted with appropriate safety precautions so they are not exposed to cyanide gas released when the acid is added.

Toxicity Characteristic Leaching Procedure

The Toxicity Characteristic Leaching Procedure (TCLP) (SW-846 Method 1311) is used to assess whether or not a material can be classified as a hazardous waste. This is a lab analysis used on both organic and inorganic substances. The test was developed when waste was being placed in "dumps." Initially, dumps were unlined, un-engineered landfills through which water could infiltrate. When it rained, rainwater would encounter the waste, pick up contaminants in the waste, and flow out of the dump, potentially impacting groundwater, surface water, etc.

The EPA developed TCLP to simulate this reality. They first determined, using toxicology, concentrations they considered hazardous if found leaching from a landfill. Then, in a lab, they would take waste and put it in a container with an acidic solution, either acetic acid or a mixture of acetic acid and sodium hydroxide. The analysis requires the solution to have a pH of 4.93 to simulate the acidic characteristic of landfill leachate. As stated above, acid causes contaminants to be released from waste more so than with neutral water. The solid waste and acidic solution are placed in a container that is rotated for approximately 18 hours to simulate water's travel through a landfill. The solid waste is removed and the solution is analyzed to determine the concentration of the contaminants that "leached out" from the waste. Theoretically, this extracted leachate would represent what could potentially migrate into groundwater.

To explain further the purpose of this test, let's take a soil contaminated with lead as an example. The soil (a solid) sample may have a lead concentration of 1,000 mg/kg (ppm). In many cases, this soil would have to be removed and hauled to a landfill for disposal due to the high lead concentration. However, you first have to determine into which type of landfill is should be disposed. If the material is hazardous, it goes to a hazardous waste landfill (Subtitle C landfill), which is more strongly engineered than a municipal waste (Subtitle D landfill). If the soil is non-hazardous, it goes to a municipal solid waste landfill; these are also engineered quite well. To determine if the sample is hazardous, you will run the TCLP analysis. If that soil, with a total lead concentration of 1,000 mg/kg, leaches 5.0 mg/l or more into the acid solution, the soil is considered hazardous and it goes to the Subtitle C landfill. If the lead concentration in the solution is less than 5.0 mg/l, then the solid waste is non-hazardous and it goes to a Subtitle D landfill. It matters how much of that contaminant will "come out" or leach from the solid waste and into the acid, not the concentration of the solid waste. A key point in this is that you submit a solid expecting a result in mg/kg but you get a result in mg/l. It is not the solid you are analyzing, but the contaminants that come out of it when it is in an acidic solution.

Table 2.8 presents the contaminants and the maximum concentrations for TCLP analysis, including select VOCs, SVOCs, metals (RCRA 8), and pesticides.

A similar analysis, called the Synthetic Precipitation Leaching Procedure (SPLP) (SW-846 Method 1312), has been developed to simulate acid rain infiltrating a contaminated site. Again, contaminants are leached from a solid in an acidic solution that represents acid rain. An interesting component of this procedure is that if your site is west of the Mississippi River, the acid solution has a pH of 5.0. If your site is east of the Mississippi, the solution has a pH of 4.2, which is more acidic than the TCLP analysis. The pH is more acidic in the East to represent the nature of the precipitation that falls in that area.

Wet Chemistry and Miscellaneous Analyses

Groundwater and surface water samples are often analyzed for water quality parameters, and are often grouped into the Wet Chemistry category. These parameters define not only water quality for living organisms, but they may also suggest conditions that help when assessing potential remediation solutions or future sampling events. They may also suggest conditions beyond the obvious, for example high alkalinity could suggest more than a high pH, it could suggest the presence of calcium. Please note that these analyses are not limited to water and most can be run on soil, as well.

Table 2.8. Toxicity Characteristic Leaching Procedure (TCLP)

Metals	EPA Haz Waste Number	CAS Number	Regulatory Limit (mg/l)
Aresenic	D004	7440-38-2	5.0
Barium	D005	7440-39-3	100.0
Cadmium	D006	7440-43-9	1.0
Chromium	D007	7440-47-3	5.0
Lead	D008	7440-92-1	5.0
Mercury	D009	7439-97-6	0.2
Selenium	D010	7782-49-2	1.0
Silver	D011	7440-22-4	5.0
Volatiles			
Benzene	D018	71-43-2	0.5
Carbon Tetrachloride	D019	56-23-5	0.5
Chlorobenzene	D021	108-90-7	100.0
Chloroform	D022	67-66-3	6.0
1,2-Dichlorobenzene	D028	107-06-2	0.5
1,1-Dichloroethylene	D029	75-35-4	0.7
Methyl ethyl Ketone	D035	78-93-3	200.0
Tetrachloroethylene	D039	127-18-4	0.7
Trichloroethylene	D040	79-01-6	0.5
Vinyl Chloride	D043	75-01-4	0.2
Semi-Volatiles			
2-Methylphenol	D023	95-48-7	200.0
3/4-Methodphenol	D024	108-39-4	200.0
p-cresol	D025	106-44-5	200.0
1,4-Dichlorobenzene	D027	106-46-7	7.5
Hexachlorobenzene	D032	118-74-1	0.13
Hexachlorobutadiene	D033	87-68-3	0.5
Hexachloroethane	D034	67-72-1	3.0
2,4-Dinitrotoluene	D030	121-14-7	0.13
Nitrobenzene	D036	98-95-3	2.0
Pentachlorophenol	D037	87-86-5	100.0
Pyridine	D038	110-86-1	5.0
2,4,5-Trichlorophenol	D041	95-95-4	400.0
2,4,6-Trichlorophenol	D042	88-06-2	2.0
Pesticides & Herbicides			
Chlordane	D020	57-74-9	0.03
2,4-D	D016	94-75-9	10.0
Endrin	D012	72-20-8	0.02
Heptachlor	D031	76-44-8	0.008
Lindane	D013	58-89-9	0.4
Methoxychlor	D014	72-43-5	10.0
Toxaphene	D015	8001-35-2	0.5
2,4,5-TP	D017	93-72-1	1.0

Alkalinity—Total alkalinity is a measurement of the amount of acid (e.g., sulfuric acid) needed to bring a sample to a pH of 4.2. At this pH, all the alkaline compounds in the sample are "used up." Alkalinity is reported as milligrams per liter of calcium carbonate (mg/L $CaCO_3$). Alkalinity is influenced by geology, salts, certain plant activities, and certain industrial wastewater discharges. High alkalinity may cause one to consider what could be in the water or soil that is causing the abnormal condition, like the presence of calcium, magnesium, or strontium.

Ammonia, nitrogen—The presence of ammonia (NH_3) in groundwater suggests an anaerobic condition, meaning lack of available oxygen. In such conditions, nitrogen is converted from relatively harmless NO_2 or NO_3 to NH_3.

Biochemical Oxygen Demand (BOD)—Wastewater from sewage treatment plants often contains organic materials that are decomposed by microorganisms, which use oxygen in the process. The amount of oxygen consumed by these organisms in breaking down the waste is known as the biochemical oxygen demand or BOD.

Chemical Oxygen Demand (COD)—Chemical oxygen demand (COD) does not differentiate between biologically available and inert organic matter and it is a measure of the total amount of oxygen required to oxidize all organic material into carbon dioxide and water. COD values are always greater than BOD values, but COD measurements can be made in a few hours while BOD measurements take five days.

Chloride, Chlorite—Both are by-products of chlorination of drinking water. Elevated concentrations may cause anemia in infants and young children.

Conductivity—Conductivity is the ability of water to pass an electrical current. It is affected by the presence of dissolved solids such as chloride, nitrate, sulfate, and phosphate anions or sodium, magnesium, calcium, iron, and aluminum cations. Conductivity is also affected by temperature: the warmer the water, the higher the conductivity therefore, conductivity is reported at 25 degrees Celsius (25 C).

Dissolved Oxygen (DO)—The amount of dissolved oxygen present in water is crucial to aquatic flora and fauna, and is also important regarding chemical processes. Fish, for example, require about 5 ppm DO in water—if it falls less than that you can get fish kills. Also, if DO is absent, an anaerobic or reduction condition exists, which affects other chemicals. For example, with sufficient DO, nitrogen oxidizes to nitrate (NO_3), which is mostly harmless. In anaerobic conditions, nitrogen reduces to ammonia (NH_3), which is harmful. Likewise with sulfur, with sufficient DO it oxidizes to sulfate (HSO_4), which is harmless at lower concentrations. However, in anaerobic conditions, sulfur reduces to hydrogen sulfide (H_2S), which is foul-smelling (rotten egg odor) and toxic. Without DO, methane (CH_4) reduces from carbon rather than oxidizing to CO_2. And finally, iron reduces to iron II (Fe^{2+}), causing iron stains, and oxidizes to (Fe^{3+}). Consider swamp gas, which is produced in anaerobic conditions. It contains methane and hydrogen sulfide and has a byproduct of iron pyrite (FeS_2).

Nitrate, Nitrite, Nitrogen—Nitrogen and its oxides are often found in fertilizers. Excessive amounts of nitrates and nitrites in surface water will promote algae growth and eutrophication, which, in turn, will cause a decrease in DO. In groundwater, nitrate and nitrite indicate aerobic conditions, but could also suggest that run-off from fertilized areas are impacting groundwater. In addition, nitrate and nitrite in drinking water can cause harm to infants who drink the water. Sources of nitrate include runoff from fertilizer use, faulty septic tanks, and sewage.

Perchlorate (ClO_4)—Most of the perchlorate manufactured in the United States is used as the primary ingredient of solid rocket propellant. Wastes from the manufacturer and improper disposal of perchlorate-containing chemicals are increasingly being discovered in soil and water.

pH—A measurement to determine if a sample is acidic (<7) or basic (>7). Water usually has a pH between 6.5 and 8.0.

Phosphorous, total—This is a nutrient to plant growth and, when found in surface water, it can promote algae growth, resulting in low DO. In groundwater, phosphorus indicates aerobic conditions. Its sources include run-off from fertilized lawns and cropland, faulty septic systems, and runoff from animal manure storage areas.

Sulfide, Sulfate—Bacteria can convert sulfide into sulfate, which is a black slimy substance. You may note this substance on bailers or tubing that have been pulled from a monitoring well with such conditions.

Total Dissolved Solids (TDS)—TDS consist of particles that will pass through a filter with pores of around 2 microns (0.002 cm) in size. These particles may include calcium, chlorides, nitrate, phosphorus, sulfur, iron, and other metals.

Total Organic Carbon (TOC)—This analysis is typically run on soils to assess the concentration of carbon in soil. Carbon inherently attracts many contaminants and serves as a cleanser of soils. For this reason, a site with a higher concentration of TOC in the soil can be left with increased concentrations of contaminants in the soil, allowing that the carbon will "clean" those contaminants.

Total Suspended Solids (TSS)—TSS are particles that will not pass through a 2-micron filter. They include silt and clay particles, plankton, algae, fine organic debris, and other particulate matter. Many contaminants, including pesticides and metals, adhere to suspended solids and affect analytical results. For this reason, you often collect two samples for metals—one filtered and one unfiltered. The unfiltered sample will have higher metal concentrations than the filtered. This is important in that you might not have a groundwater contamination problem but rather a TSS problem, which could be solved through physical filtering rather than chemical treatment. The need to filter a sample is often based upon the turbidity of the sample.

Turbidity—Turbidity is a measurement of light passing through water and is measured in nephelometric turbidity units or NTUs. Meters can measure turbidity over a range from 0 to 1000 NTUs. A clear mountain stream might have a turbidity of around 1 NTU, whereas a large river like the Mississippi might have a turbidity of around 10 NTUs. While the turbidity is not a measurement of suspended solids, the presence of suspended solids will increase the turbidity. If the turbidity in a sample is too high (greater than 5 NTU), you should filter the samples prior to analysis.

Analytical Methods—SW-846

Throughout this chapter I have mentioned SW-846. You will typically read in your Workplan and Field Sampling Plan that all samples you collect will be collected and analyzed using methods in SW-846. SW-846 is an EPA publication entitled *Test Methods for Evaluating Solid Waste, Physical/Chemical Methods.* It is the EPA's official collection of analytical and sampling methods that have been evaluated and approved for use in complying with RCRA regulations. It is a guidance document providing acceptable, although not required, methods for use in responding to RCRA-related sampling and analysis requirements. While compiled for RCRA, it is used throughout the sampling and analytical community.

SW-846 was first issued in 1980 and has changed over time as new information and data were developed. Advances in analytical instrumentation and techniques are continually reviewed by EPA's Office of Solid Waste (OSW) and incorporated into periodic updates. The current manual is Edition IIIA and contains approximately 3500 pages. The majority of the methods are designed for labs to analyze properly the samples submitted to them, and provide a standard methodology to facilitate consistency between labs. They are complex, but contain nuggets that will help you better understand your role in the field. SW-846 is available online at *http://www.epa.gov/epaoswer/hazwaste/test/main.htm.*

3

Groundwater Sampling

Groundwater sampling seems to many to be the most burdensome and complex type of sampling. It requires an abundance of equipment, sample containers, and patience. When compared to surface soil sampling, which is simply digging a hole and putting the soil in a jar, groundwater sampling is rather complex. You have various equipment to use in numerous wells, you have back up equipment, you have to monitor conditions of the water (pH, temp, level, etc), and you have numerous sample containers that are for several analytes, some with preservatives some without. In addition, you have the continual risk of cross-contamination. The process requires a great deal of preparation, which begins in the office.

Office Preparation

It is essential to read the site-specific Work Plan or the Field Sampling Plan before beginning the office preparation. These plans often contain requirements well above what would normally be done at a site. They should be reviewed carefully prior to the project.

Once you have the plan in mind, and the reasoning behind it, you are ready to begin. Start in the office by gathering equipment and supplies for the sampling effort. A rule of thumb for this effort is 15 percent of your time on the whole project will be spent in the office, preparing for your field work and cleaning up after the job. Some project managers forget this aspect of field work when they budget. They think a week in the field will be two people at ten hours a day for five days or 100 man-hours when it is actually 115 hours. You should allow a day for preparation and a half-day for clean-up. Remember this in your budgets. Too many people have to "eat" hours because the work was not properly budgeted. Good office preparation can help to reduce field time by maximizing efficiency.

The office preparation includes many tasks. First, you will have to order sufficient sample containers for the job. Once they arrive, the amount of work you do with them in the office may vary. Some people like to fully label each sample container for each sample to be collected, and then organize the containers in coolers such that when they get to MW-1, all the containers needed for that well are in one cooler, already labeled. The only items to fill in after this preparation are the time sampled and sampler's initials. My preference was to label the containers to an extent that did not include the sample number, that way I was free to use any container I could find. Another thing you could do in the office with the sample containers is add the preservative, if it has not already been added by the lab. Again, this could be done in the field, but preparing in the office saves time. If you are on a project that uses "freeze-packs" rather than ice, make sure you put a sufficient number of them in the freezer well enough in advance so they serve their purpose.

Office preparation includes obtaining and calibrating your water quality instruments. Do not wait until the day before you sample to take this step. If something is not working properly, you need sufficient time to repair or order a new unit. For the pH-conductivity meter, ensure that you have enough buffer and calibration solution, and make sure you have spare batteries.

Make sure to test all equipment to ensure that it functions properly. This includes equipment such as pumps, generators, meters, and vehicles. Ensure that each has enough fuel, oil, and other fluids that allow it to operate. Be sure to buy rope, baggies, and other supplies that may be purchased at the local hardware store.

Print the forms you need for the field, including borehole logs, monitoring well development forms, and sample collection forms. Make sure you have keys to the wells. A good idea is to have each lock keyed-alike, meaning that one key will open all locks. It makes matters easier if all of your sites are keyed alike to one key. This will ultimately save on the use of bolt cutters and locks.

Upon return from the field, time must be taken to ensure the equipment is decontaminated, working, and put away. If a piece of equipment is not working, then it should be repaired with the time and cost charged to that job—the project manager can decide if the client should pay for it. If you have depleted the equipment shelves for your job, you should replenish them and charge the time and supplies to the job.

Field Work and Equipment

Once you have prepared everything you need to do in the office, it is time to begin the field work. If you have prepared properly in the office, your field work will go remarkably smoothly.

Water Level Indicator

There are a few aspects of the water level indicator that you should keep in mind. First, it does not work in distilled water, only in water that has ions in it.

Electricity does not conduct through pure water, so the water level indicator cannot transmit through it.

When decontaminating a water level indicator, do not spray the probe or the tape when it is on the spool or reel. Water can penetrate into the electronics and damage the meter. Always un-reel the tape into a bucket to do the decontaminating.

When you take a water level, you have a round well with 360° of "sides" from which to measure. Often, the well riser is cut unevenly and measurements from one side can differ from another side, which could result in water "flowing uphill." There are three things you can do to avoid uneven measurements. First, if you are present when the well is installed, insist to the drillers that they should cut it evenly. Second, regardless of the even nature of the riser (as opposed to the uneven nature of a driller), always measure water levels from the north side of the riser to provide consistency. Finally, use a file or knife to put a notch in the north side of the riser to indicate from where water levels should be measured. If the riser is stainless steel, use a file. Remember to inform the survey crew when they get the elevation of the riser to look for the notch, or to get the elevation from the north side. If the surveyors do not follow these guidelines, all of your measurements will be invalid. Do not use a Sharpie® to mark the measuring point of the riser because they have VOCs and it is best to keep them away from monitoring wells.

If you are trying to assess the direction of groundwater flow by water levels, take a round of water levels before you start sampling. This is especially important on large projects that may last several days or weeks. Water levels can fluctuate throughout the day due to pressure differences in the atmosphere. As a weather-front approaches the site, the water levels may rise or fall. So, if you measure a few wells in the morning and a few in the afternoon after the weather has changed, the levels may not accurately reflect the direction of groundwater flow.

To assure accurate water level readings, it may be necessary to allow the level to equilibrate after you remove the cap to the well. Putting a cap on a well is like putting your thumb on a straw in a glass of water. Once your thumb seals in the water at a certain level, you can raise or lower the water level in the glass without affecting the level in the straw. When you release your thumb, the level in the straw will equilibrate to the level in the glass, just like in a confined aquifer. Removing the cap allows the water level in the well to equilibrate with the level in the aquifer. I have seen water levels fluctuate six to ten inches upon removing a cap, which makes a huge difference when determining groundwater flow. Take your time and let the water equilibrate. If your measurements result in the groundwater flowing north when it is actually flowing east, the whole remediation effort will be for naught.

If you have forgotten the water level indicator, there is a device you can make called a "popper" that can be used to get the levels. It may be crude, but it works. To make a "popper," buy a 1 ½" PVC cap, stainless steel wire, and an "eye hook". Screw the eye hook into the top of the cap and, using the wire, secure it to a 100-foot tape that has been decontaminated. Measure the distance between the bottom of the cap and the bottom of the tape. Drop the tape down the well and listen for the "popping" sound that occurs when the cap hits the water. You should then take your measurement and add the distance from the cap to the tape.

When you are back at the first well to be sampled, start at the clean wells and proceed to the dirty. Also, go from up gradient to down gradient to minimize the chance for cross-contamination. Be sure to determine your groundwater levels in that order, too.

Bailers

Bailers are generally fabricated from four materials; stainless steel, Teflon®, PVC, and polyethylene. Each has its advantages and disadvantages. Stainless steel bailers are durable and can be used repeatedly after decontaminating. They sink quickly (a characteristic of steel), which allows you to collect water more quickly than with the non-steel bailers. Also, they don't absorb contaminants and can be easily decontaminated. A disadvantage to the stainless steel bailer is that you are unable to see water or any free-product layers in the bailer. Another disadvantage is when allowed to drop down the well, there is turbulence in the water that may allow VOCs to volatilize. This turbulence may cause TSS interferences with metals analyses, as well, so the analytical data may not represent that which is in the water. However, an experienced sampler should be able to pick up the bailing rhythm, estimate when the bailer will hit the water surface and stop the freefall action, then let the bailer sink softly into the water. Before I knew of these finer sampling points I was a very fast bailer. On one site, I was able to bail just over 0.5 gallons per minute in a 60-foot well (water column at 50') using a stainless steel bailer. It is a rate I have not come close to matching when using a non-steel bailer. In hindsight, I should not have bailed with such rapidity. The site was heavily contaminated so I do not think the samples were too adversely affected.

The Teflon® bailer can be considered disposable, re-usable, and dedicated. The quality you buy may dictate the number of times you use it. Because it's Teflon®, it can be decontaminated. These bailers are used on sensitive sites that require Teflon® equipment. They are sensitive in that they want to avoid any absorbance of the contaminants into the sampling devices, thereby affecting the analytical results—contaminants do not absorb into Teflon®. Teflon® can be either translucent or opaque, so take note when ordering. Obviously, you can see product layers through the translucent. A caveat with Teflon® bailers is that some of them are constructed with thin walls to minimize cost, but then they are flimsy. On several occasions I have held a sample container in one hand and grasped the middle of the bailer in the other hand. When attempting to pour the contents into the sample container, the bailer folded in my hand causing a dropped sample and an awkward situation. I recommend buying thicker-walled bailers.

The polyethylene (PE) bailer is usually disposable, although they can be dedicated to a well. They are far less expensive than stainless steel or Teflon®. Another advantage to the PE bailer is that they are translucent, allowing you to see if there are free-product layers in your water column. A disadvantage to the PE bailer is that some compounds may absorb into the material, causing inaccurate analytical results. Likewise with the Teflon® bailer, buy the thicker walled product. Finally, PE bailers can have weights in them. Some will have one, either at the top or bottom, and others have two. I prefer the two-weight bailer because it sinks faster, thereby speeding up the sampling process.

The PVC bailer is heavier than Teflon® and PE but lighter than steel. It is opaque, so you cannot see product in it. It should be dedicated to a well and not decontaminated and reused.

With respect to sampling with a bailer, there are some who will not use bailers to sample for VOCs, stating there is too much turbulence during collection and that it affects the quality of the sample. This not only occurs when dipping the bailer into the water, but also as you pull the bailer up and when you pour the water into the voa vial. Keep in mind, as you tilt a bailer to pour from the top, more water is exposed to air and there is an increased chance for volatilization. The proper method for VOC sample collection with a bailer is to insert a device in the bottom of the bailer that allows water to pour from the bottom, thereby minimizing air exposure.

The Rope and Knot

The rope I prefer for bailing is white braided nylon. Braided nylon rope does not remember its shape when packaged, unlike the yellow polyethylene rope which has a "memory" regarding the shape when it is packaged. The yellow poly rope will always try to bend where it was folded in the package or on the spool. This is very frustrating when you are trying to bail with long arm movements to collect the rope. I typically use ¼-inch or ⅜-inch diameter rope.

When using a bailer to develop a well or sample groundwater, it is important to keep the bailer attached to the rope you are using. It is a dreadful feeling when that bailer drops and there could be financial or water quality implications with a lost bailer.

Most bailers are three or four feet long, but I recommend the three–foot long variety, as they are easier to manage. If one is down a well, you can no longer collect DNAPLs from the well. As DNAPLs, by definition, sink to the bottom of an aquifer, the dropped bailer prevents you from effectively sampling the bottom. So, a dropped bailer that cannot be retrieved where DNAPLs had to be sampled would mean you have to drill and install a new well so the base of the aquifer could be accessed. I once dropped a stainless steel bailer down a well. The well did not have to be re-installed, but there was concern over the eventual decomposition of the bailer and its affects on water quality (e.g., increased amounts of iron in water).

Figures 3.1 through 3.9 demonstrate the method for tying the knot that I prefer to use. I have never had one of those knots come loose on me, they are easy to tie, and the harder you tug on the rope and bailer, the tighter the knot becomes. I first learned this method by the description, "rabbit running around a tree three times."

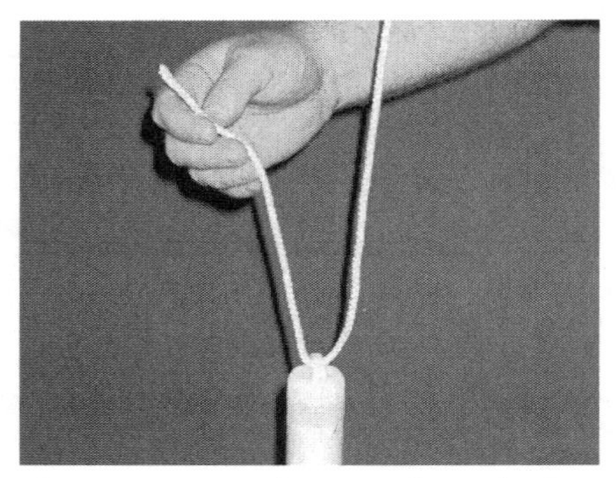

Figure 3.1. Put the rope through the "eye" of the bailer.

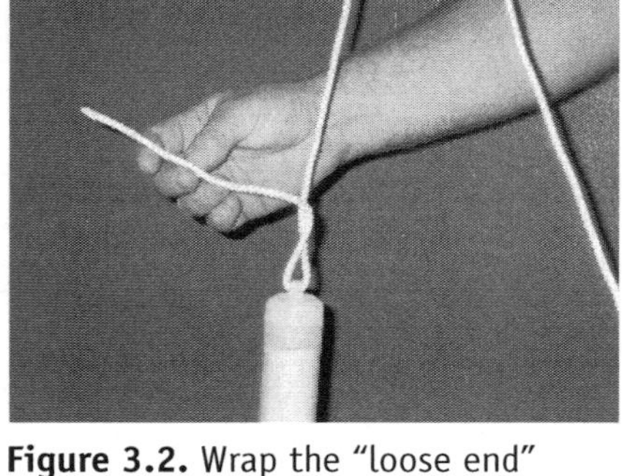

Figure 3.2. Wrap the "loose end" around the longer portion.

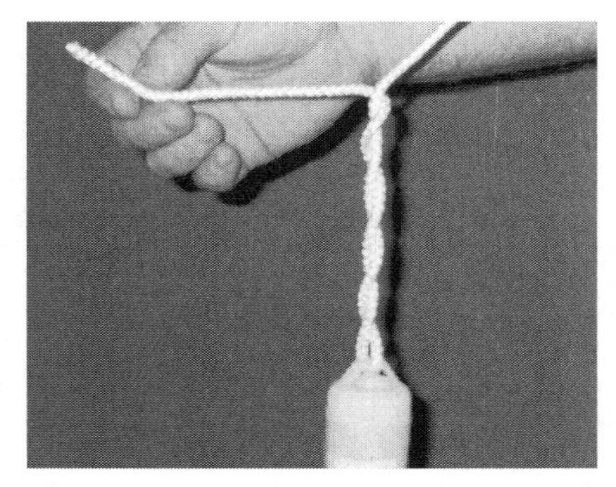

Figure 3.3. Continue wrapping around three or four times (this was explained to me as, "the bunny chases around the tree three times - hey, it works).

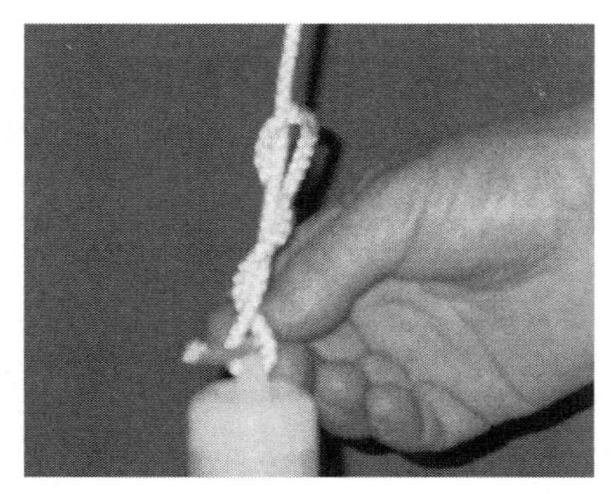

Figure 3.4. Bring the loose end down and put it through the open area above the "eye of the bailer."

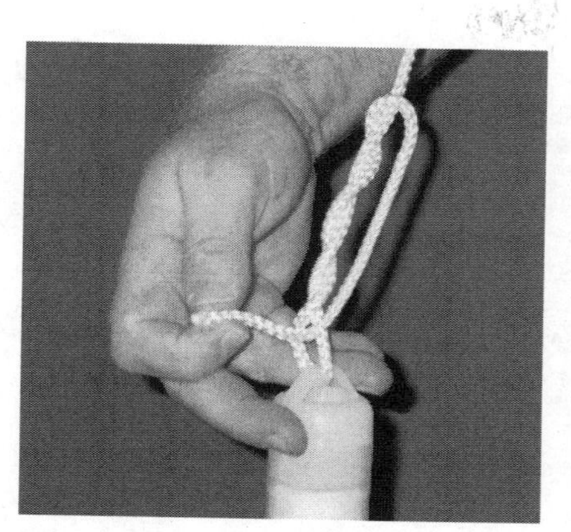

Figure 3.5. Grasp the loose end and start to pull it through.

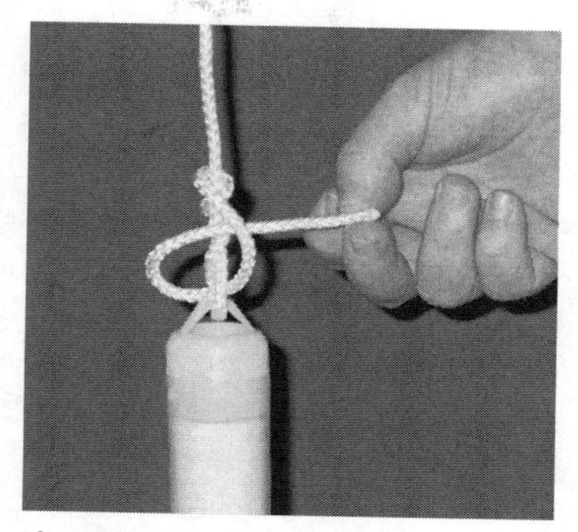

Figure 3.6. Grab the loose end and long end and pull tightly.

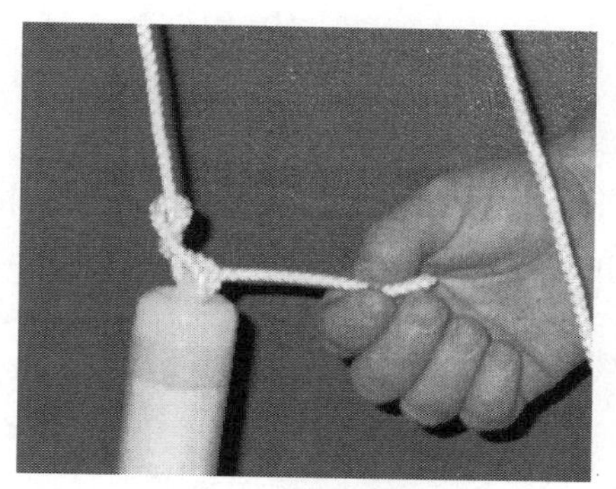

Figure 3.7. As you pull, the wrap around area will shrink down toward the "eye of the bailer."

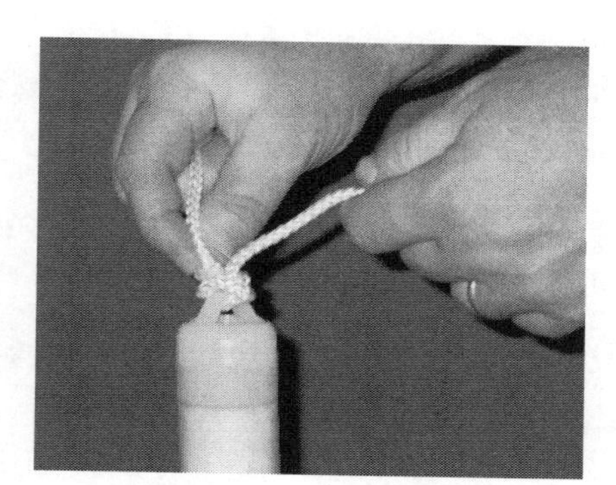

Figure 3.8. Push the wrapped area down to the "eye of the bailer."

Figure 3.9. Pull both ends tightly. This knot should suffice. The harder you pull on the rope, the tighter the knot becomes.

Retrieving a Bailer

If you choose not to heed this knot advice, there may come a time that you will have to retrieve a bailer. The standard method requires using a fishing line with hooks and literally trying to fish for the bailer or the rope. Standard issue in many groundwater tool kits is fishing line, hooks, and lead weights (sinkers). Please do not ever put a lead weight into a monitoring well, regardless of how secure it is to the fishing line. If the line breaks, you have a lead source in the well. This is not the sort of thing you want to tell either your project manager or the client, and the method is highly unsuccessful.

I prefer retrieving a bailer by using ¾-inch threaded PVC pipe. This pipe comes in five-foot lengths and can be found at well-supply stores. On one piece, drill a ¼-inch hole at the bottom. Buy several hooks of varying shapes and sizes, the kinds that have "eyes" for screws are good. You may have to try a sewing and notions department or an electrical department. If you can find stainless steel pieces, all the better. Also, buy some stainless steel wire. Now, affix the hooks to the PVC pipe in whatever manner you can, either by screwing them in place or using the wire, wrap and tie the hooks in place. Loop the wire through the ¼-inch hole to make it more secure. Try to have the hooks pointing in any and every direction. With the hooks firmly attached, lower the pipe into the well and add more sections of PVC until you reach the lost bailer. When this is accomplished, twist and turn the PVC so as to hook onto either the rope or the eye at the top of the bailer. This should take but a moment. When you have it, pull it up and out of the well.

Pumps

Submersible Pump

Submersible pumps can be fast when in the well but cumbersome in preparation. Some submersibles can produce up to 9 gallons per minute (gpm), depending on depth, which is much quicker than my 0.5 gpm with a bailer. With submersible pumps, you need an energy source, typically a 110 volt or 220 volt generator, although, there are some units that require compressed gas to operate (e.g., nitrogen). In either case, there is a lot of equipment to carry to the field when compared to a bailer and rope. Another disadvantage is that they can be difficult to decontaminate. There are many moving parts, and some are sensitive to the solvents often used for decontaminating. Some submersible pumps have impellers in them and they can be worn down in silty or sandy water. If you are using a pump with impellers in such an environment, be sure to have spare impellers on hand. They can be easily replaced, provided you have The Ideal Tool Kit (see Chapter 8). As with all electrical equipment, make sure you have shrink wrap for wire repairs, and wire cutters. The more complex the system, the more can break down.

A key advantage to submersible pumps is that they can get a lot of water out of a well in a hurry. Nine gpm is lightening fast. Consider filling a five-gallon bucket in about 30 seconds. When you have considerable volume to remove, it is convenient to have a fast pump. These pumps usually have a control mechanism that can adjust the rate of removal, thus allowing you to collect VOCs without the water spraying everywhere.

A suction pump I have used with considerable success is the Whale pump. It is about six inches high and 1 ½ inches in diameter. It runs off a 12 volt car battery and sometimes comes either with about 30 feet of lead wire. They make dual pumps, as well, that have about 60 feet of lead wire. The single can pump 3.0 gpm while the dual can pump 2.8 gpm at greater depth. To use the pump, simply add whatever length of tubing you need, put the pump down the well and attach the wires (battery clamps) to a battery.

While they sell duals now, in the past I have taken up to three single pumps, popped off the turquoise bottom, and put three inline by using about a one-inch piece of tubing between each. I added a length of wire as needed, using shrink wrap to join the wires, and I have pumped out an 85-foot well. If I recall correctly, I got about 0.25 gpm, which was quicker than bailing.

The interesting thing about Whale pumps is that they were designed for the boating industry to pump potable water from storage tanks to faucets or showers. You may be able to buy the pumps cheaper at a marine shop than an environmental shop.

Suction Pump

Suction pumps are not often used for sampling, although I have used them for developing wells. The one that I have used frequently is commonly called a trash pump or garbage pump, and can be found at your local equipment rental store. Its purpose is to move a lot of water and sediment in a short time frame. They are especially effective in wells with a lot of sediment. It is very difficult to plug one of these pumps. The pump has a two-inch intake and two-inch discharge. For a two-inch well, I would reduce the intake side down to about a ½ inch, using fittings found at a hardware store and attach polyethylene tubing. On the discharge side I would use the standard issue two-inch flexible hose that comes with the unit. However, if I had to collect the development water, I would reduce the discharge down to a garden hose size which could be placed into a drum or bucket.

A trash pump is gasoline-powered, so you have to take precautions not to contaminate your equipment and the well when using the unit. Also, this unit has to be primed, meaning you have to add water to a vessel within the pump in order to get suction. A caveat for this pump is if the pump stops for any reason, you have to immediately pull the intake side out of the well. If you do not, the water in the pump will drain back into the well and, since you cannot effectively clean the inside of these pumps, cross-contamination is possible. Also, if you have a garden hose in a drum for water collection, when the pump shuts, off the water in the drum will be siphoned back into the well. You cannot leave the area for any reason when you are using this pump.

The trash pump is effective only to a depth of 20 feet. Also, it is a burdensome system for wells to which you cannot drive.

Bladder Pump

Bladder pumps are ideal for low flow and/or deep well uses. They use a compressed air source, either a compressor or a nitrogen bottle, to squeeze a bladder in the pump which is down the well. The squeeze and release cycle causes the water to come to the surface. These pumps can be used to great depth (400 feet) and come in sizes that will fit in a ¾-inch well. The drawback is the amount of equipment needed to support the unit. If you use a compressor, you may have to bring a generator.

Inertia Pump

In order to use the Inertia pump, place a foot-valve at the base of a piece of tubing, and jerk up and down until the water comes out of the top of the tubing. You may attach a motorized mechanism to do the up and down jerking motion to create a more complex system. This system does not work consistently, so if it fails, have The Ideal Tool Kit handy. I have not found this method effective for sampling because of the up and down motion. However, it is effective for well development, especially in wells with a lot of fines. If you develop with this pump you can sample with a bailer or some other device.

The manual method is very good for remote places or small diameter wells, as foot valves are made for ¾-inch wells. The mechanized version is more suited to two-inch wells to which you can drive.

Peristaltic Pump

This pump is used primarily for purging, sampling, or filtering samples. It has a very low flow rate so it should not be used for well development except, perhaps, for very small diameter wells, e.g., ¾ inch. This pump requires a 12 volt electric source, and most come with battery clamps and/or cigarette lighter adaptor. These pumps are quite reliable and rugged. They have a dial to control flow rate so you can purge at a high flow rate and collect VOCs at a low flow rate. Be sure to decrease the flow for VOC samples because at high flow rates the peristaltic action spurts the water out of the tubing causing it to splash when it hits a hard surface like a sample container. If you are using the tubing in the peristaltic pump repeatedly, be sure to decontaminate it properly. Decontaminate the outside of it, too because when you handle the tubing you could cross-contaminate. There is often a mindset to skimp on the length of tubing you use because it is relatively expensive (approximately $2 per foot). While I agree you should not use six feet when two feet will suffice, I encourage you to use a liberal amount. The increased labor costs in trying to contort a short piece to do something a long piece would do with ease may outweigh the cost of the additional tubing.

Water Quality Equipment

The primary pieces of water quality equipment used during sampling are the thermometer, pH meter, and conductivity meter. They are often a combined unit and require a sample to be collected during the purging or developing event, and the probe is to be placed in the sample. A digital readout is provided. Additional water quality equipment includes a turbidity meter, dissolved oxygen meter, and oxygen-reduction potential (ORP) meter. With the exception of a turbidity meter, you can often get the ORP, DO, pH, conductivity, and thermometer in one flow-though cell. For these, you hook up the tubing from the well to the cell and, as the water flows, the readings are provided in "real time." For the ORP meter, the readings are either negative mV or positive mV. If a negative reading is given, the water is in a reducing state (little dissolved oxygen). If it is positive, the water is in an oxidizing state.

The purpose of this equipment is two-fold. First, it tells you the obvious parameter that it is measuring. Second, this equipment is used to indicate if the water you are removing from a well has stabilized, meaning that you are removing water that has consistently similar water quality parameters. As you remove water, if the pH is jumping from 6.5 to 7.0 to 6.3 to 6.8, then the water is not "the same" each time. If it is not the same then you should not be sampling it because it is not representative of stabilized conditions in the groundwater. Once the parameters stabilize, then you are consistently removing "the same" water. The guidelines for stabilization differ from one parameter to the next. The following presents the criteria used to indicate stabilized conditions for most water quality parameters:

Stabilization Criteria[1]

Dissolved Oxygen	±0.3 mg/l
Turbidity	±10% (for samples greater than 10NTUs)
Specific Conductivity	±3%
ORP	±10 mV
pH	±0.1 unit
Temperature	±0.1°C

These criteria apply to averages taken over three readings. For example, if your previous three temperatures were 16.5, 16.3, and 16.6 degrees Celsius (average 16.5) and your next sample was 16.7°C, you would have to continue purging because the difference is greater than 0.1°C.

Well Development and Purging

Well development and purging are similar in concept, as both involve removing a certain volume of water from a well. However, you develop a well upon completing its construction and you purge a well before sampling. There are common methods and unique techniques for each.

Development

A well is developed upon completion to remove water and sediments associated with drilling and well construction. During the drilling process, water is often added for numerous reasons, including to keep the borehole open or to provide lubrication to the augers or rods. Although the water is potable, since it is not part of the aquifer it has to be removed. Also, drilling, by nature, stirs up sediments from the geological formation and they remain in the well upon completion. Other sediments may be introduced from the sand pack used around the screen, which might contain fines. Well development has to remove this foreign matter which is not part of the naturally occurring groundwater.

Development equipment typically includes pumps and bailers, but may be augmented by surge blocks or compressed air. The pumps and bailers are quite obvious in their intended purpose, but, their use will vary depending upon site conditions. As stated previously, submersible pumps for two-inch wells are not suited to silty groundwater conditions since the impellers wear out quickly. Other pumps and bailers are better suited to well development. If you use a pump, you should use the highest flow rate you can obtain to try to draw down the water

[1] From the U.S. EPA's *Groundwater Sampling Guidelines for Superfund and RCRA Project Managers.*

level in the well then allow recharge. This seems to be the best method for development.

Surge blocks or plungers are devices placed into a well at a certain depth and create a surging motion which causes fines to loosen from the filter pack into the well. The fines are removed with a pump or bailer. The surge block operates under a concept similar to that of a toilet plunger. Because the surge block seals the well, pushing on it in a plunging motion causes water and pressure to loosen fines. If the well is set in geology with fines, this practice should not be used because you can loosen excessive fines from the formation.

Surging can be conducted with pressurized air as well. For this method, an air line is lowered to the surge block and initiated. The air pushes the water out through the screen and filter pack, loosening fines. A variation of this is to lower a line with a metal stinger into a well, then hook the line to a nitrogen bottle and turn on the air. The pressure will clear out water and sediments, causing the material to jettison from the well. This is especially useful for wells that have a lot of sediment in them, and the bottom is not accessible due to siltation. A caveat for this method is that it is used on geotechnical wells, not environmental. Using a pipe with a 90° angle, you can jet air through the well screen to loosen fines. This is especially useful for freeing mud from mud rotary drilling. Surging and jetting methods should not be used within 48 hours of completing well construction because those methods might adversely affect the grout.

Based upon my research, you should develop a well until you have removed three well volumes or the water quality data has stabilized. The well volume is determined using the following calculation:

$$\text{Well Volume (V)} = \pi r^2 h * (7.48) \quad \text{[Equation 1]}$$

where:

π ≈ 3.14

r = radius of monitoring well in feet (ft)–if you have a two-inch diameter well, the radius is one-inch. Divide one by 12 to get the radius in feet.

h = height of the water column in ft.

7.48 = the gallons of water in one cubic foot.

Monitoring well diameters are typically 2-, 3-, 4-, or 6-inches. If you have a two-inch well, there are 0.1631 gallons per foot. If the water column is 27 feet, then one well volume is 4.4 gallons.

Well diam. (inch.)	2	3	4	6
Volume (gal/ft)	0.1631	0.3670	0.6524	1.4680

If you do not want to do the math, Figure 3.9 presents a graph of well volumes based upon various diameters and water column heights. Simply go to the well diameter, follow the line to your water column height, and you have your well volume.

Thinking back on experience, I recall developing wells until I removed 10 borehole volumes, which was our standard protocol. I had to calculate the volume in the two-inch monitoring well and add to it the volume of water in the six-inch borehole, accounting for porosity of the filter pack. This is frustrating because we were in the field three times longer than we had to be.

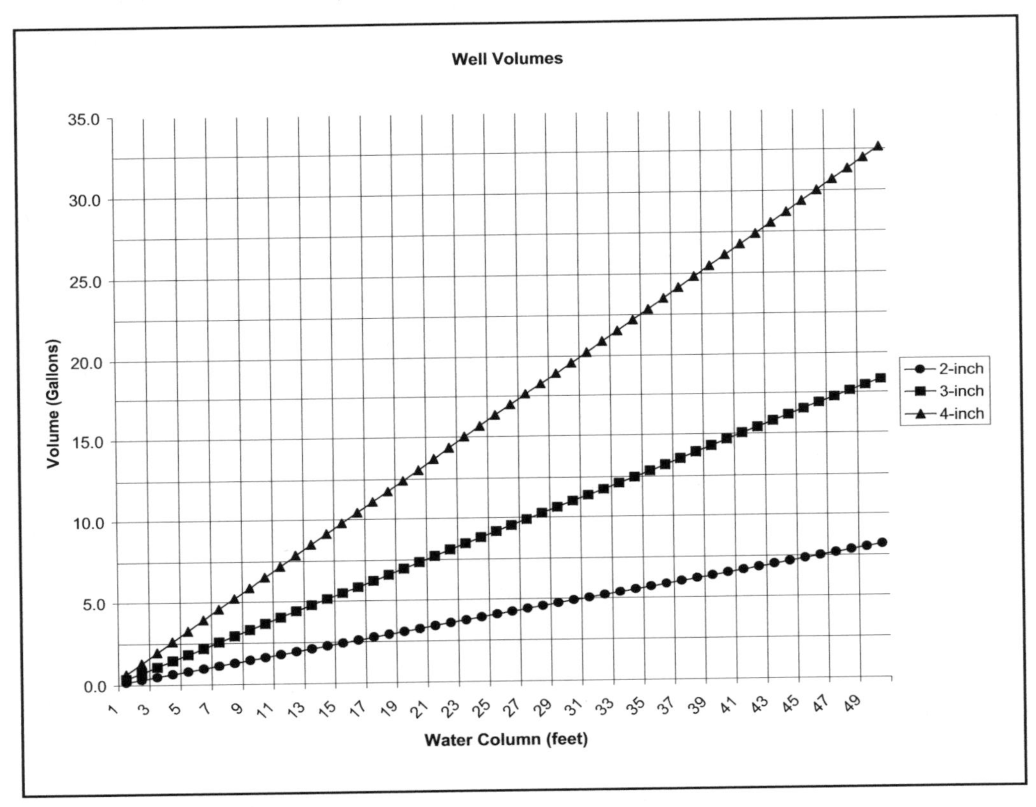

Figure 3.10. Well Volumes

Purging

Purging is conducted immediately prior to sampling to assure that the water being sampled is representative of existing water conditions in the aquifer. You cannot just get to a well and collect the sample because the water might be stagnant or foreign matter might have been introduced from the top of the well, either purposely or accidentally. There are two methods used for purging: one is the "well-volume method," and the other is the "micro-purge method."

Well-Volume Method

Like regular development, the general rule of thumb for this purging method is to remove three well volumes. EPA guidance documents say three to five well volumes is better. You can use the same methods as development for purging. If the well is slow to recharge, meaning you can remove all the water quickly and it does not refill quickly, you can sample after removing one well volume. The water that recharges will be fresh from the surrounding water formation. Otherwise, you have to remove the three volumes. However, it is not just based upon volume; you measure water quality while purging and cannot sample until those have stabilized.

Also, with this method you have to make sure the bailer gets deep into the well screen area, to the depth from which you want to collect the sample. However, you do not want to get so deep that you disturb sediment at the bottom of the well, thereby affecting your turbidity readings, unless you're sampling DNAPLS.

Micro-Purge Method

The theory behind this method is that if you pump slowly from a single location within the well screen, you will clear out the existing water without disturbing the stagnant water that usually rests above the well screen. Research has demonstrated that pumping a small volume of water at a slow rate will remove sufficient water to present stabilized water quality. This method is believed to result in samples that are more representative of existing conditions in the water formation. While the equipment can be costly, especially when compared with a bailer, often the equipment will be dedicated, resulting in savings with equipment blanks, set-up, etc.

Another advantage and cost savings is that the volume removed is quite low. It is not dependant on well volumes, but you should remove at least three equipment volumes, including the tubing, flow-through cell, and pump in order to determine water quality stabilization. Keep in mind that 50 feet of 3/8" diameter tubing holds about 0.28 gallons, flow-through cells hold about a quart, and a pump might hold $^{1}/_{10}{}^{th}$ of a gallon. At flow rates recommended between 0.1 and 0.5 liters per minute, one volume could take about one minute. The time to purge, therefore, is minimized and there are savings on water disposal (if applicable) or handling of purge water.

The Sampling Event

Before you get started, place plastic sheeting around the well. Things drop, rope will touch the ground, and caps will fall from your fingers, so put down the sheeting so they do not touch potentially contaminated surfaces. When you begin by filling sample containers, start with the VOA vials. Most labs require two vials to be filled, although some prefer three. This holds true regardless of your sampling method. If you are using a pump that has flow-rate control, decrease the flow rate before you fill the vials. As you know, a VOA vial cannot have any headspace, meaning there can be no bubbles in the vial once the cap is on. Here are a few tricks for filling these vials so there are no bubbles in them.

1. When pouring or pumping the sample into the vial, tilt it like you are pouring a carbonated beverage and trying to avoid creating foam. The concept works the same with the vial.
2. Make sure you fill the vial with a convex meniscus. If you pour gently, you can get about a ¼ inch of water rising above the top of the vial. I like to fill the vial cap with water, to provide some water that I can pour even more gently to top off the vial. Make sure you rinse the cap to prevent cross-contamination.
3. If there are a few bubbles sitting at the top of the meniscus, touch the edge of the vial cap to the bubbles and move the cap to the glass rim of the vial. Capillary action will cause the bubbles to go the side and/or stick to the cap, allowing you to put the cap on.
4. If you are sampling in a formation with a lot of suspended solids, and the solids are from limestone or dolomite geologic formations, the solids might react with the HCl preservative and cause excessive effervescing. If this happens, discard the sample and rinse out the vial so there is no HCl. Then, refill the vial. Be sure to note on the Chain-of-Custody that there is no preservative in the vials and the turnaround time is now seven days (for more information on this see Chapter 8 – Valid Data).
5. EPA guidelines suggest that you invert the vial upon filling and capping and observe the vial for ten seconds to assess the presence of bubbles. I have always rapped the vial against my palm to cause bubbles to rise into view when observing.
6. Keep in mind that even a tiny bubble can collect volatilized gas from the sample. When the lab takes the aliquot for their analysis, it would be minus any gas that has volatilized into the bubble or headspace. When measuring to parts per billion, it does not require a large bubble to affect the results. If you can see the bubble, it obviously occupies more than a billionth of the vial volume and, therefore, can collect more than a billionth of the VOCs within the sample.
7. An EPA guidance document I read stated that VOA vials, upon collection, should be placed on their sides in the cooler. I have never done

this, nor do I know anyone who has. The lab personnel to whom I have spoken have never seen this done, nor do they recommend it. If the cap is not tight, water could leak out if the vial is on its side.

After you collect VOCs, the sample collection order is as follows: SVOCs, Metals, PCBs/Pesticides/Herbicides, and Water Quality Parameters. If you have to filter metals and you are sampling with a bailer, fill a glass measuring cup, decontaminated, of course, with your sample. Then, use a peristaltic pump or other pumping device with the appropriate filter (typically 45 μm) on it to transfer the sample into the sample container. I had a friend who was filtering landfill leachate samples and did not realize how quickly a filter could become plugged. When the filter plugged, the hose broke away from the filter and began spraying leachate into his face. Change the filters often when pumping an oozy substance to avoid a similar fate, and wear your safety glasses.

Once you fill the sample containers, put them on ice immediately. Before you dismantle your equipment and leave the well, double check that all the samples have been collected and are secure in the cooler. This means physically looking at each label to assure that each sample, for each parameter, is in the cooler. Make sure any duplicates, MS/MSDs, or blanks have been collected.

When you are sure that you are finished, then dismantle and decontaminate the equipment. While one decontaminates, the other can prepare for the next location by labeling sample containers, calculating well volume (based upon previously obtained water level), checking supplies, recalibrating equipment as necessary, or completing the Chain-of-Custody forms for the last sample.

4

Soil Sampling

There are two primary methods for collecting soil samples: from a drill rig or from an excavation. However, the actual methods and reasons for putting soil in a sample container are similar for each collection method, and the tools used are often interchangeable. In this chapter I will discuss various tools and methods used to collect samples. The pros and cons of each will be considered, and tips to assist you in your sampling efforts will be identified.

Tools

When collecting a sample using a hand shovel, make sure that the soil you collect does not come in contact with the steel. Likewise in large excavations (test pits, tank excavations), when using a back-hoe or excavator, the soil collected should not come in contact with the bucket. The steel used in hand shovels and excavators is porous and can collect contaminants. Even decontaminating will not always remove contaminants from a steel shovel. In addition, garden tools are not recommended for sampling purposes, as they are often chrome-plated. For hand excavations, dig to the approximate depth of sample collection. Then, using a stainless steel tool like a trowel or scoop, scrape away the final layer of soil to get to the collection zone. You should then decontaminate the stainless steel tool before collecting the sample because it touched the soil that touched the hand shovel. It seems pedantic, but when analyzing to parts per billion, it is the precaution that needs to be taken.

When at a large excavation, have the back-hoe remove a bucket of material from the collection zone and bring it up to ground surface. The open face of soil should not have come in contact with the bucket, itself. Collect your sample from within that face, making sure to dig in a couple of inches to get a sample that was not exposed to the elements.

Plastic tools will also absorb contaminants and, in some cases, release chemicals from their composition. As numerous contaminants of concern are used for manufacturing plastic, using plastic to collect a sample defies logic. The only utensils used should be constructed of stainless steel, glass, or Teflon®. The EPA says "equipment constructed of plastic or polyvinyl chloride (PVC) should NOT be used to collect samples for trace organic compound analyses" (EPA Contract

Laboratory Program (CLP) Guidance for Field Samplers), which implies that they can be used for collecting highly contaminated samples. If the sample is obviously contaminated, like oozing with a substance, then perhaps plastic would be acceptable. Personally, I would not take the chance. I would not want to be on the witness stand saying I sampled with plastic when the analytical results are being questioned.

Supplies

Stainless steel sampling equipment can be purchased at places like Ben Meadows and Forestry Supply. However, you may find yourself needing something that cannot wait overnight for delivery. A great source for stainless steel tools is your local high-end kitchen supply store. I have purchased items at Williams-Sonoma™ and other kitchen stores. They have knives, mixing bowls, scoops, trowels, and various other implements. On one job I had to dip into an oil reservoir to sample oil. The opening was quite narrow so I ended up using a stainless steel coffee scoop purchased at Williams-Sonoma. I bent up the handle so I could use it for dipping, and it worked.

You want to avoid cheaper kitchen items because the steel may be plated, which can chip or peel, thereby exposing the underlying metal. Also, restaurant supply stores, which may not be as ubiquitous as the aforementioned chain store, have an abundance of stainless steel supplies. Restaurants must use stainless steel in their kitchens for health reasons. Keep in mind that stainless steel is expensive, but it is cheaper than re-sampling. Think in the long term. A $50 scoop or bowl is cheap when compared to several thousand dollars.

Borehole Samples

When you have good, experienced drillers, the samples will seem like they are flying at you. Some drillers take delight when they are faster than you. You have to delight in knowing that the borehole is logged accurately and samples are collected efficiently. If that takes time, so be it. If the drillers give you a hard time, you are under no obligation to hire them again. Having said that, a good set of drillers with whom you work comfortably and efficiently are worth their weight in gold. I recommend buying the coffee each morning. In one area I worked it was a tacit rule that the consultant buys the coffee. When learning that lesson I had to leave site to buy coffee in order to deal civilly with the drillers. Drillers can make your life easier or more difficult and when you are just starting out in this business, you may not recognize when they are making it rough on you.

When the sample comes to you, make sure the driller tells you top from bottom. This may not be obvious from geology, especially as the samples come from greater depths. Also, get the driller into a pattern of always placing the sample in the orientation you like, either top to left or top to right. It makes more sense to me to have it left to right since that is the way we read. The driller will open the split spoon or cut open the GeoProbe tube for you. Place your tape measure (preferably the wooden engineer's rule) next to the sample with 0' at the top. Quickly you can see the recovery length and note that on your borehole log.

Make sure you are ready to collect a sample for headspace analysis. If you are not ready, ask the driller not to open up the sample. The longer it sits open, the more volatilization will occur. As soon as it is opened, collect a sample for headspace. It should be taken either from the area that appears to be most contaminated or from the full length of the sample retrieved. Passing an organic vapor meter (OVM) over the length of the sample will indicate if there is an area of high contamination.

Headspace Analysis

The headspace analysis is the measurement of VOCs emanating from a soil sample that is in a sealed sample container. The measurement is intended to be taken in the air (space) at the top (head) of the container to where volatiles would rise, hence the name headspace. The headspace is measured with an OVM (also known as a photo-ionization detector, or PID).

I have seen several ways to collect a headspace sample. Perhaps the most laborious, yet most proper, is to put the soil in a Mason jar, cover the top with tin foil, and screw the lid on without the standard middle piece of a Mason jar lid. When you are ready to measure the headspace, you can push the OVM probe through the tinfoil to measure. This to me seems the most accurate method because you read the air (or space) at the top (the head) of the jar and, if you are careful with the probe, no volatiles will escape when you push the probe through the foil. The drawback to this method is that the jars are expensive, and time consuming if you want to decontaminate them to be used again. In addition, you have to label the jar with the boring and depth of sample. In all cases, using a magic marker or Sharpie® to label when measuring for VOCs may cause interference with the reading. You cannot write on glass and expect it to remain. I think the best way to label is write the information with a ballpoint pen on a piece of paper and place it in the jar before you insert the sample. Alternatively, you could put a piece of duct tape on the side and write the sample ID on the tape.

The more common headspace analysis method I have seen is to place the soil sample in a plastic Ziploc® bag and puncture the plastic with the probe to measure the headspace. The disadvantage to this is that it is difficult to pinpoint the head in a plastic bag that is lying flat. Once you pick up the bag, the headspace location changes so when you poke the plastic you may be getting the "sidespace." If you open the zipper and put in the probe, some volatiles may escape from the opening. This is an inexpensive method for headspace collection, which is why it is more common. Also, bags take up less room than other containers. Be careful when labeling the bag because volatiles from the marker may seep through the plastic and affect the readings.

Finally, another method I have seen is putting soil samples in a plastic container with a lid. To measure the headspace, slightly lift the lid or slice the lid with a knife and slip the probe into the container. The disadvantage is that some volatiles may escape as you lift the lid and slip in the probe. The advantage to this method is the container can be labeled easily and either disposed of or decontaminated.

In all cases you should put the soil in the container and let it stand at room temperature, ideally, for a while to allow for volatilization. In cooler temperatures, you should put the samples in your running car to allow the samples to heat up to room temperature. A good and legitimate excuse to warm yourself is to get the headspace readings.

When collecting headspace analyses, you should either take a representative portion of the material from the length to be sampled or collect from the area that appears to be most contaminated. Your project manager should provide guidance on headspace location. If you are collecting two-foot split spoons, take material from the full length of the sample. Likewise, from a four-foot GeoProbe®, scrape material from two two-foot lengths. Put this material into your chosen headspace container and let it sit. If you are sampling from a test pit or other excavation, scrape material from the side wall at the approximate depth from which you are collecting samples.

Sample Description

When a sample comes from a wet boring, there will be a thin coating of mud over it—be sure to scrape it off so you can identify accurately the underlying geology. Also, when you collect your sample, be sure that none of the mud coating remains. It may carry contaminants not related to the soil you are about to sample. You will need to scrape it off before you take your headspace sample, too. A dry, dusty coating may be present on samples, as well, especially when boring through gravel, sand, or concrete. In all cases, when you collect your sample, make sure it is representative of that layer and not cross-contaminated. Describe the samples using guidance presented in Chapter 2.

Method 5035

In the past, when a soil sample was collected for VOC analysis, we would stuff as much soil into a four-ounce sample container as we could. When the sample got to the lab, the technician would scoop out something from roughly the center of the container and use it as the sample aliquot. That aliquot may have been the most contaminated piece in the container or it may have been the least. We may have wanted a specific portion of the sample analyzed but we had to completely fill the container with soil to minimize headspace, which was difficult with stiff silty clay. The technician had no idea from which of the container he should take the aliquot.

Now we have SW-846 Method 5035: Closed System Purge-and-Trap and Extraction for Volatile Organics in Soil and Waste Samples. This method allows you to take three five-gram aliquots from the sample matrix and have the lab analyze that which you want analyzed. It is not random, it is based on your knowledge in the field. This knowledge is based upon your visual and olfactory senses and OVM readings.

When you observe the sample matrix, whether it is from a drill rig, GeoProbe or in a hole in the ground, you have to assess which area you want to sample. Typically one will sample from the location with the most apparent contamination within the two-foot sample interval. That could be based upon presence of free-product, staining, or discoloration, or it could be based upon strong odors or elevated OVM readings. On many projects, only the sample within the borehole with the highest OVM reading would be submitted for analysis. Before 5035, I would fill a four-ounce container in each interval and after I took the headspace readings I would submit the designated sample. It was quick, inexpensive, and easy. Now, however, we have to buy the 5035 sample set ($8 to $12, depending on method used) and collect at each interval. Each 20-foot boring adds over $100 to the project. The alternative is to guess where you think the most contamination will be and just collect a sample from that interval.

There are several approved ways of collecting samples in Method 5035, but the most common methods seem to be using the EnCore™ sampler or using plastic syringes and VOA vials. The EnCore sampler is a polymer vessel that will hold approximately five-grams of soil (a 25-gram model is available, as well). The sampler is placed in the base of a stainless-steel T handle then the T handle is pushed into the soil until the sampler is full of soil. The sampler is removed from soil, and the T handle and a polymer cap that seals the soil is placed on the sampler. The sampler is placed in a zipper package, labeled, and placed in a cooler at 4°C. For each sample location, two samplers are required. However, if you expect VOCs high concentrations (i.e., greater than 200 µg/kg), then a third sampler should be filled. The samplers must be sent to a lab and the soil removed within 48 hours. The lab will either analyze them immediately or place them in preservatives, such as sodium bisulfate and methanol. The advantage to the EnCore sampler is that you do not have to work with pre-weighed sample vials in the field. The disadvantage is that the sample has to be preserved within 48 hours. This requires pre-planning when sampling on a Friday or Saturday or near holiday weekends.

The other method commonly used involves plastic syringes, and syringe holder, and preserved VOA vials, which are provided by your lab. The syringe is placed in the holder then pushed into the soil. The syringe holder is designed to allow the syringe to hold approximately five-grams of soil. The syringe is then placed over a VOA vial and the soil is pushed into the vial. A sample set from the lab typically contains three vials; two having a five ml solution of sodium bisulfate and water, and one having approximately five ml methanol. Also, each vial contains a stirring bar. The sodium bisulfate is used to bring the pH of the sample to <2, which prevents biological degradation in the sample. The methanol will also minimize biological degradation but will not decrease the pH to less than 2.0. The sample placed in methanol will be used for analysis if the concentrations are greater than 200 µg/ml. Each vial is weighed in the lab and sent to the field. Therefore, it is important not to add a label to the vial and not to spill any of the preservative. Any change in weight beyond the addition of soil will affect the sample results. Be attentive when placing soil in the sodium bisulfate. Some soil will cause effervescence that could cause the vial to overflow, changing the weight.

An advantage to the syringe method is that it is generally less expensive than the EnCore. The initial outlay for the EnCore sampler and T handle is higher, but over the long run it could be less expensive. With the syringe method, the samples are preserved immediately so there is no concern with getting the samples to the lab in a timely manner. However, spilling the preservative or changing the weight in any way besides adding the soil will affect the analytical results. Prior to going into the field, be sure to check with the regulatory agency as to what method they prefer.

Compositing Samples

The purpose of compositing is to get a general idea of the contamination at a site either laterally or vertically. For example, if you wanted to determine the approximate contamination across the top two-feet of a site, you could collect ten samples from the top two feet, analyze them then average the ten results. Or, a less expensive way would be to composite the ten samples from across the site and analyze one sample to represent the average for the site. Likewise, if you wanted to approximate the contamination vertically, you could composite samples at different intervals and analyze the one sample rather than averaging several samples.

You cannot composite samples for VOC analysis. The mixing process to composite will volatilize the compounds and the results would be inaccurate. Collect the VOC sample then you can composite for the rest. Compositing is simply mixing together all the soil of a sample interval then putting it in sample containers. Typically you composite in two-foot intervals (each two-foot split spoon sample), but longer intervals can be composited. After taking the VOC aliquot, put the rest of the material into a stainless steel bowl and break up larger chunks of soil and mix it together using a stainless steel spoon or trowel. Put the mixture into your sample containers. Be sure to decontaminate the mixing tools between samples.

The risk in compositing is that you may be diluting the samples as you composite, thereby providing a more positive picture of the site than may actually exist. Conversely, a high concentration in a discrete area could cause you to think a large area is contaminated. In addition, acceptance of compositing by state agencies varies considerably across the country, and when allowed, might be restricted in application.

5

Surface Water, Sediment, and Miscellaneous Sampling

For the typical sampler, surface water, sediment, and other miscellaneous types of sampling such as drum sampling, PCB wipes, etc., are not encountered as frequently as soil, air, and groundwater sampling. Granted, there are those whose sole job is to sample surface water, but you can only sample it if there is surface water at your site. Chances are that you will have soil, groundwater, and air at your site. These other types of sampling are not as prevalent as the "big three." The following chapter gives a brief summary of the common methods and equipment used for these other media.

Surface Water

An important thing to remember when you are sampling shallow surface water in a stream, river, creek, or any moving water, is that you must to start downstream and work upstream. If you start upstream, the sediment that stirs from your efforts will affect the samples you collect down the river. This is less of an issue if you are sampling a deep river with little or no chance of touching the bottom. However, if you do not know the depth and you lower a sampling tool, then you will not know when or if it will hit the bottom.

Another consideration is the presence of strata within the water body. The goals of the project might require certain depths or a certain water quality to be sampled. Before sampling occurs, water quality data should be collected from the bottom to the top, in one meter (39 inch) intervals. Data should include temperature, pH, and DO, but could also include ORP. There is monitoring equipment available (e.g., Hydrolab) that can be lowered to obtain real-time data. There is no need to collect water samples then measure water quality. Once the strata have been identified, you can proceed with sample collection. With that in mind, let's look at some sampling methods.

Direct Method

This method involves lowering a sample container, using your hands, into the water. Typically you are walking in the water when you get this sample. In my experience, these samples are usually collected below the surface, as opposed to collecting the water on the surface. The surface can carry contaminants or impurities that do not necessarily affect water conditions or represent water conditions at the time of sampling. If you open the container and invert it then lower it into the water, the surface water will not enter into the container. When you get to the desired depth, turn the bottle into the stream to collect the water. You have to be downstream of the container for this method. Again, if you stand upstream, you will disturb the bottom, which will affect the sample.

You should not use sample containers with preservative, as there is a chance that the preservative will be washed out of the container. It is better to fill the container, then add the preservative. As this is not likely for VOCs, be sure to note on the Chain-of-Custody form that the sample is not preserved, thereby decreasing the holding time from 14 to seven days (see Chapter 8 for more information).

An advantage of this method over others is that you do not have to decontaminate any equipment, which saves time and money. A disadvantage is walking in a stream will stir up sediment and may affect the water sample and any sediment samples you need. Also, walking on stream bottoms can be precarious and waterproof boots or hip-waders often are not completely waterproof.

Sampling Equipment

The Kemmerer Bottle is a Teflon®, acrylic, or stainless steel tube attached to a rope, and has stoppers at each end. It is best used from structures such as bridges and piers or from boats. It is lowered vertically into the water with the stoppers open at each end. The stoppers are secured open by a trigger mechanism. The bottle is lowered to the desired depth and a "messenger" is sent down the rope, hitting the trigger. The messenger is a heavy brass weight that slides down the rope. The trigger causes the stoppers to close, thereby containing the sample. The bottle is pulled up, and water is transferred to a sample container.

With this method you can use sample containers with preservative in them. A drawback is that you have to decontaminate the bottle between samples and collect equipment blanks.

Similar equipment includes the Bacon Bomb (sample thief) and Van Dorn sampler. Each piece of equipment is lowered by a rope to the desired depth. The Bacon Bomb is opened by pulling a second line, then releasing when the bomb is full. There are Bacon Bombs that are triggered when they hit the bottom, but those are used primarily to collect samples in storage tanks. The Van Dorn has stoppers open on the way down. When it is at the sample depth, it is pulled sideways to trigger the stoppers closed.

Perhaps the easiest sampling equipment to use is the Dip Sampler. It allows you to stand on the stream bank and lower a container into the water. Upon retrieval, the contents are transferred into the appropriate sample container.

Sediment Sampling

The sediment is considered the mineral and organic layer that lies beneath an aqueous layer, like water. The water might be flowing, as in a river or stream, or it could be static, as in ponds, lakes, and impoundments. If you are collecting surface water and sediment samples at the same location, be sure to collect the water first, then the sediment. With each sampling method, when the material is removed from the water, carefully decant the water and place the remaining material in the sample container. When decanting, make sure that you do not lose the fines within the sediment. As each of these methods requires equipment, decontaminating is necessary and equipment blanks might be required.

Sampling Equipment

In shallow, slow-moving water, the most basic method of collecting sediment is to use a hand-held scoop or trowel. You can collect it either standing in the water or from the edge, if you can reach. If you are sampling while standing in the water, be sure to stand downstream of the collecting point so as to minimize sediment disturbance. The scoop or other hand-held equipment should be stainless steel. The effective depth for this method is approximately six inches.

Bucket and tube augers can be used to collect samples at greater depth. The two are used in tandem to reach greater depths. The bucket auger is wide and has a larger capacity, while the tube auger is long and narrow. Both can accommodate a T handle that measures about five feet but extensions can be added. For this collection method, the bucket auger is used to dig a pre-hole for the tube auger. When the bucket auger has reached the top of the interval to be sampled, the tube auger is then placed in the larger hole and pushed into the sediment, which is then retrieved. Sample recovery can be increased if the holes are augered at a slight angle so the material does not fall out of the tube auger when it is being removed.

The Eckman dredge is a lightweight sampler used for soft shallow sediments. It is often lowered into a water body by a rope, but it also can accommodate a T handle for shallow water assignments. To use, the Eckman dredge is lowered slowly into the water to within approximately six inches of the sediment, at which point it is dropped. The dredge is closed by releasing a triggering messenger. The dredge is then raised slowly, and the sediment is removed and placed in sample containers.

The Ponar Dredge is considerably larger and heavier than the Eckman, thereby allowing greater penetration into the sediments. The large Ponar weighs approximately 50 pounds while the Eckman weighs approximately 10 pounds. The Ponar is lowered gently with a rope or steel cable with winch until it is near the sediment. It is then released in a free-fall into the sediment. When the dredge hits the bottom, the "jolt" triggers the device to close. An issue with this equipment is that if it is lowered too quickly, when it hits the water surface it could close. This is a frustrating prospect at great heights and without a winch. When used properly, the Ponar can collect larger sized particles and a larger volume than the Eckman.

Drum Sampling

Drum sampling is fairly simple when you know generally what is in the drum and you can get into the drum. Opening a drum is beyond the scope of this book. However, how not to get into the drum would include the following method, used at what is now a Superfund site: the site owners would line up drums of unknown substances and, at a reasonably safe distance, shoot them to see if they would blow up. Some did, while others would just drain into the ground. There are numerous pieces of equipment that can be used to open a drum, so choose a safe one. I should note that the Ideal Tool Kit does not include a brass universal bung wrench. If you do considerable work with drums, you should get one. Once the drum is open, there are two primary methods for sampling drum contents.

Glass Thief

The glass thief is the most common and least expensive method for sampling drums. It is a glass tube and operates under the same premise of a straw in a glass. Think back to your physics experiments when you would put the straw in the milk, put your thumb over the top, and lift the straw from the cup. You would then move the straw to another vessel and remove your thumb, thus releasing the milk. The glass thief operates under the same physical premises. When you're finished, you can either safely break the glass thief and put it into the drum or attempt to place it into the drum without breaking it. Get your project manager's approval for either of these methods so that a glass thief does not interfere with future drum disposal plans.

Coliwasa

The Coliwasa, short for composite liquid waste sampler, is a long tube with a rod in the center and a neoprene stopper at its base. The bottom stopper can remain open as the Coliwasa is lowered into the drum. When it reaches the bottom, you can push down on the tube to close the stopper then lift up the sample. The Coliwasa is good for getting representative samples from the entire drum. The samples will not be overly disturbed and you can likely identify stratification. It is beneficial for multi-phase samples. A disadvantage to this sampler is that it can be difficult to decontaminate, but must be decontaminated between samples.

PCB Wipes

There are complete guidance documents issued by the EPA on PBC sampling. Their document called "Verification of PCB Spill Cleanup by Sampling and Analysis" (Document EPA-560/5-85-026) can be found at their web site, *www.epa.gov.* The document goes into great detail on the PCB sampling and should be followed. For general PCB sampling you still put the sample media in a sample container, but I would like to briefly discuss some finer points about PCB wipes, as this sampling is rather unique.

PCB wipes are intended to assess the PCB (or other constituent) concentration on smooth, non-porous surfaces such as painted surfaces and steel. It was designed to imitate a person placing their hand on or leaning against a surface and coming in contact with PCBs. The wipe test involves using a cotton gauze or similar substance, which has been soaked in hexane, and is wiped against the surface. The hexane is used to simulate a sweaty hand. The area wiped is 100 cm^2 or roughly 16 in^2. Have a look at the palm of your hand, it measures roughly 100 cm^2 or 4-by-4-inches.

The best way to measure the area is to use a pre-cut template. Some labs provide these cardboard templates as part of their PCB analysis service. Otherwise, you can measure by placing masking tape in a 100 cm^2 pattern (10 cm by 10 cm). The shape of the area does not matter as long as it measures 100 cm^2. Actually, that is the preferred size. If the area to be wiped is smaller than 100 cm^2, you can use a smaller template, but wipe results are typically reported in µg/100cm^2. If you use a smaller area, you have to inform the lab so they can report the results per the smaller area or adjust the results to represent 100 cm^2.

When you wipe the template, you wipe back and front, and then side to side, to make sure you get the full area. It is not necessary to scrub the gauze into the surface, as you are not trying to clean the surface of all the grime. You are simply simulating a lean against the wall or some other incidental contact.

When you get the results back, they will be in µg/unit area. PCB regulations dictate remedial and disposal options based upon parts per million (ppm). For example, if something has a PCB concentration of =50 ppm and <500 ppm, it is PCB-contaminated, and therefore has certain remediation and disposal options.

If the PCB concentration is = 500 ppm, then the disposal options are limited and more costly.

As stated above, the wipe results are reported in µg/unit area. The EPA has stated that a wipe sample having a concentration = 10 µg/100 cm^2 and < 100 µg/100 cm^2 is the same as being PCB-contaminated or having a concentration = 50 ppm and < 500 ppm.

PCB regulations are found at 40 CFR 761, which is also called the Mega-Rule. The regulations are difficult to understand, but repeated reading and application will help. Also, the EPA is typically eager to answer questions to clarify these regulations.

6

Air Sampling

by Richard Trzupek

Types of Sampling

Air samples, like most environmental samples, are taken for one of two basic reasons: to gather internal data, or to attempt to prove compliance with a regulation.

In the first case, experience and complex equipment is often not needed. You can approximate the amounts of many air pollutants with relatively simple equipment and a common sense approach. Such sampling is usually, though not always, enough to aid you in the decision-making process.

Internal data gathering, also known as "informal testing," is used to tell you if you might have a problem, or definitely do not. It is used to screen potential pollutants and concentrations, to find direction. The end result of informal testing is usually either, "I'm sure we're okay," or "we need to do more research; there may be a problem here."

"Formal testing," on the other hand, must follow strict and complicated methodologies published by the U.S. EPA. The equipment is expensive and the techniques are complex. With a few exceptions, most formal compliance testing is contracted to outside, professional stack-testing firms.

In this chapter, the focus will be placed on the tools, equipment, and techniques used for informal testing. Whether you are screening ambient air for specific pollutants at a waste site, measuring contaminants in the work place, or attempting to approximate emission rates in a smoke stack, this is the form of air sampling one is most likely to engage in outside of a compliance demonstration.

Tools of the Trade

Air sampling tools come in two types:

1) Continuous analyzers, which determine a parameter virtually instantaneously, and
2) Bulk samples, which gather a sample for subsequent, off-site analysis.

Common continuous analyzers include Photo-Ionization Detectors (PID), Flame-Ionization Detectors (FID), gaseous analyzers, hot-wire anemometers, and Lower Explosive Limit (LEL) detectors.

PIDs are primarily used to screen concentration of organic compounds at remediation sites and in the ambient air. They are most sensitive to typical gasoline compounds and other organics that are classified as "aromatic" or "olefinic." They are much less sensitive to straight-chain, "aliphatic" compounds.

FIDS are the organic tools of choice for higher concentration streams, like those in the stack of a painting or printing line for example. They are sensitive to all organics, but trickier to use.

PIDs are usually self-contained and relatively light. FIDs require an outside source of fuel, hydrogen, or a hydrogen/helium mix.

The detectors of each are prone to contamination, especially dust, or a large amount of water. Some PIDs and FIDs have pre-filters which can obviate the problem. If not, the use of a simple filter on the end of the sample probe is highly recommended.

Gaseous analyzers utilize a variety of chemical principles to determine the concentration of compounds like oxygen, carbon monoxide, nitrogen oxides, and sulfur dioxide. They range in size from portable to bulky.

Interferants can affect some analyzers, especially the less expensive models. Likewise, temperature fluctuation can create error in some models as well. Operators are well-advised to consult their operating manuals to determine the safe range of operation.

Hot-wire anemometers are used to measure air flow. While not yet accepted by the EPA as an approved technique, modern units, when used according to instructions, compare quite well to the approved methods.

It is impossible to over emphasize the importance of a good air flow measurement. When you attempt to determine the mass emission rate of a compound (lbs/hr, for example), the final calculation involves multiplying a pollutant concentration by the air flow. The best concentration measurement in the world can be horribly skewed by an incorrect air flow determination.

LEL detectors are primarily a simple, reliable, safety tool. They are used to screen potentially explosive atmospheres. They should not be used to measure organic concentrations. That function is much better left to PIDs and FIDs.

Bulk sampling techniques utilize "sampling trains," which are multi-component assemblies designed to condition and collect samples for analysis. Common bulk sampling trains include Tedlar bag trains, SUMMA canisters, impinger trains, and filter trains.

Tedlar bags are a simple, effective means of collecting gaseous samples when the target pollutant concentration is relatively high, about 10 ppmv in most cases. A Tedlar bag train usually consists of a stainless steel or Teflon® sample probe, Teflon sample line, a Teflon-headed pump, Teflon line, and the Tedlar bag. A moisture trap or particulate filter may be added if needed.

SUMMA canisters are self-contained sample system, usually supplied by an air quality laboratory. SUMMA canisters are used to measure low and ultra low concentrations of gaseous compounds, typically in the parts-per-billion to parts-per-trillion range.

Impinger trains are used to collect compounds in condensable concentrations or those that can be readily absorbed by a liquid in the impingers. These trains typically consist of a stainless steel or Teflon probe, Teflon line, glass impinger train, Teflon-headed sample pump, and gas rate meter. The sample pump draws the gas through the impingers when it bubbles through and may react with the liquids contained within.

Filter trains are simple, usually consisting of a probe, sample line, pre-weighted filter, pump, and gas rate meter. It should be noted, however, that particulate concentrations in the air are quite variable. A filter train can roughly approximate particulate concentrations in the ambient air, where gas velocities are basically constant. In a stack or duct with variable gas velocities, techniques that are more sophisticated are needed.

Sample Locations

The accuracy of sampling depends, to a great extent, on the choice of sample locations. Air behaves like a liquid, so choosing a calm, clean location to grab a sample will yield a much different result than if you picked a turbulent, dirty spot. The trick is to find a place that is either representative or, depending on your objective, represents a worst case.

When sampling ambient air, it is often most useful to pick a location that most closely approximates actual human exposure. Some samplers can be attached to a worker, or placed to best mimic human exposure. You must decide how close to get to the source—the nearer the source of the contamination, in general, the higher the readings will be.

Sampling air in a duct or a stack is another matter entirely. Trapped within a duct or stack, pollutants, particularly if they are particulate, can mix in uneven concentrations. As a rule of thumb, you should try to pick a sampling location approximately three-quarters of the way down a long, straight run of ductwork. The flow will be relatively calm in such a location and, as a result, the sample more representative.

Tricks of the Trade

There are, naturally, certain tricks to good air sampling that do not appear in any published EPA methodology. Some of the most valuable are as follows:

A. A leak check is an invaluable and simple means of quality assurance that should be used with any sampling equipment that utilizes a pump. The technique involves plugging the end of the sample probe and observing, usually via a gas flow rate meter, if the pump will draw a vacuum. If so, the sample train is judged to be leak-free, and therefore representative. If not, the results are unlikely to be representative.

B. Calibration gases are used to test continuous analyzers. Most often, these gases are supplied in heavy, rigid cylinders. However, as a matter of convenience, there is usually no reason to carry the heavy cylinder all the way to the sample location. A simpler and equally accurate method is to fill a Tedlar bag with the calibrated gas and use the bag to calibrate the instrument in question.

C. When Tedlar bags are used, for calibration or for sampling, the integrity of the bag is of utmost importance. The most common point of failure is the area immediately surrounding the sample valve. Accordingly, one should pick a bag supplier based primarily on how hardy their valve design is.

D. Some types of particulate will attract moisture, drawing it out of the ambient air and artificially inflating apparent particulate concentration measurements. Acid gases, especially sulfuric acid, are the most common culprit. Accordingly, the prudent technician will ascertain whether any portion of the air stream is likely to attract and retain moisture. If this condition does exist, moisture removal and location correction methods are advised.

E. Finally, the significance of accurate gas flow rate measurement was emphasized earlier in this chapter. One of the worst sources of error in such measurements is when gas flow is cyclonic or spiraling. This will usually occur at the exit of a fan or blower, or near a bend in ductwork. If these conditions are suspected, the means to determine the presence of cyclonic flow may be found in the EPA methodology.

7

Valid Data

The goal of sample collection is not to have a specimen of soil, water, or air, but to have analytical data that accurately represent existing conditions at the site. That responsibility lies heavily on the sampler. The responsibility starts before you get out in the field, stops briefly when the samples are in the lab, and resumes when the analytical data are delivered. These aspects are very important. If your samples are not collected properly, they are not valid and do not represent existing conditions at the site, rendering them useless. I have heard of a legal case where fats, oil, and grease (fog) samples were not collected properly. The data were presented in court, and because the samples were not valid, the case settlement went from several million dollars to $200,000.

Before you go into the field, you have to make sure you have the proper equipment to conduct your sampling event, and you must make sure it is clean. You also have to check that you have the right type and number of sample containers. It is best to plan ahead, before heading out into the field, essentials such as what you are going to do and what you will need. You have to order sample containers from the lab you are using. When they arrive, make sure they are what you ordered, as boxes have been known to be mislabeled. Open each box, check for the certificate ensuring that the sample containers were pre-cleaned (all containers should come with that), and check that you have the proper number and type.

Sample Preservation

Most samples collected have to be preserved in some manner to assure that the sample collected in the field maintains its characteristics until it is analyzed. Preservation includes keeping samples cold, adding an acid or a base, using amber sample containers, and analyzing within the holding times. Table 7.1 presents a list of parameters, the required sample containers for each, the preservative used, and its holding time. Each preservative has a purpose, and they are explained below.

Table 7.1. Sampling Guide

INORGANICS

PARAMETER	CONTAINER	QUANTITY (ml)	PRESERVATIVE	HOLDING TIME
Acidity	Plastic	100	Cool 4 deg C	14 days
Alkalinity	Plastic	100	Cool 4 deg C	14 days
Ammonia	Plastic	500	H2S04 to pH<2, Cool 4 deg C	28 days
BOD	Plastic	1000	Cool 4 deg C	48 hours
Bromide	Plastic	200	None	28 days
Chloride	Plastic	200	None	28 days
Chlorine	Plastic	200	Cool 4 deg C, Immediately	Analyze
COD	Plastic	100	H2S04 to pH<2, Cool 4 deg C	28 days
Color	Plastic	100	Cool 4 deg C	48 hours
Conductivity	Plastic	100	Cool 4 deg C	28 days
Cyanide, Total or Amenable	Plastic	1000	NaOH to pH>12, Cool 4 deg C	14 days
Cyanide, Reactive, pH 2	Plastic	1000	NaOH to pH>12, Cool 4 deg C	14 days
Flash Point, Closed Cup	Glass	100	Cool 4 deg C	-
Fluoride	Plastic	500	None	28 days
Hardness, Total	Plastic	100	HN03 to pH<2	6 months
Nitrite	Plastic	100	Cool 4 deg C	48 hours
Nitrate/Nitrite (waste water, chlorinated drinking water)	Plastic	200	H2S04 to pH<2, Cool 4 deg C	28 days
Nitrate/Nitrite (non-chlorinated drinking water)	Plastic	200	H2S04 to pH<2, Cool 4 deg C	14 days
Nitrate/Nitrite (non-chlorinated drinking water)	Plastic	200	Cool 4 deg C	48 hours
Nitrate/Nitrite (chlorinated drinking water)	Plastic	200	Cool 4 deg C	28 days
Oil & Grease	Glass	1000	H2S04 to pH<2, Cool 4 deg C	28 days
pH	Plastic	100	None	Analyze, Immediately
Phenols	Glass	1000	H2S04 to pH<2, Cool to 4 deg C	28 days
Phosphorus, Ortho	Plastic	200	Cool 4 deg C	48 hours
Phosphorus, Total	Plastic	200	H2S04 to pH<2, Cool to 4 deg C	28 days
Silica	Plastic	100	Cool 4 deg C	28 days
Solids, Dissolved	Plastic	100	Cool 4 deg C	7 days
Solids, Suspended	Plastic	500	Cool 4 deg C	7 days
Solids, Total	Plastic	100	Cool 4 deg C	7 days
Solids, Settleable	Plastic	500	Cool 4 deg C	48 hours

PARAMETER	CONTAINER	QUANTITY (ml)	PRESERVATIVE	HOLDING TIME
Solids, Volatile	Plastic	100	Cool 4 deg C	7 days
Sulfate	Plastic	200	Cool 4 deg C	28 days
Sulfide	Plastic	500	ZnOAc + NaOH to pH>9, Cool 4 deg C	7 days
Sulfide, Reactive pH 2	Plastic	500	ZnOAc + NaOH to pH>9, Cool 4 deg C	7 days
Sulfite	Plastic	200	None	Analyze Immediately
Surfactants, MBAS	Plastic	250	Cool 4 deg C	48 hours
Turbidity	Plastic	100	Cool 4 deg C	48 hours
METALS				
General, dissolved	Plastic	500	Filtered on site, HNO3 to pH<2	6 months
General, total	Plastic	500	HNO3 to pH<2	6 months
Chromium, hexavalent	Plastic	250	Cool 4 deg C	24 hours
Mercury	Plastic	500	HNO3 to pH<2	28 days
ORGANICS				
HPLC Pesticides (Aldicarb / Carbonfuran)	Glass vial	1000	1.2 mL Chloroacetic acid / vial, Cool 4 deg C	28 Days
EDB/DBCP	Glass vial	1000	Cool 4 deg C	28 Days
Endothall	Glass (amber)	1000	Cool 4 deg C	7 days extraction, 1 day - analysis
Pesticides and PCBs	Glass (amber)	1000	Na2S2O3 if Cl2 is present, Cool 4 deg C	7 days extraction, 40 days - analysis
Petroleum Hydrocarbons, IR	Glass (amber)	1000	H2SO4 to pH<2, Cool 4 deg C	28 days
Phenoxyacid Hebicides	Glass (amber)	1000	Na2S2O3 if Cl2 is present, Cool 4 deg C	7 days extraction, 40 days - analysis
Phthalate Esters	Glass (amber)	1000	Na2S2O3 if Cl2 is present, Cool 4 deg C	7 days extraction, 40 days - analysis
Polynuclear Aromatic Hydrocarbons	Glass (amber)	1000	Na2S2O3 if Cl2 is present, Cool 4 deg C	7 days extraction, 40 days - analysis
GC/MS Semivolatiles	Glass (amber)	1000	Na2S2O3 if Cl2 is present, Cool 4 deg C	7 days extraction, 40 days - analysis
Total Organic Carbon (TOC)	Plastic	100	H2SO4 to pH<2, Cool 4 deg C	28 days
Total Organic Halogens (TOX)	Glass (amber)	100	Na2S2O3 if Cl2 is present, H2SO4 to pH<2, Cool 4 deg C	28 days
Total Petroleum Hydrocarbons	Glass (amber)	1000	Cool 4 deg C	7 days extraction, 40 days - analysis
Volatile Organics	Glass vial	2X40	Na2S2O3 if Cl2 is present, HCl to pH<2	14 days
Volatile Aromatic Organics	Glass vial	2X40	Na2S2O3 if Cl2 is present, HCl to pH<2	14 days

Temperature

Samples are cooled to 4°C to minimize biological activity. Soil and water have microbes in them that can continue to work to "eat" the contaminants or affect the quality of the sample media. Cooling a sample minimizes, slows down, or stops their activity. While the samples are to be cooled to 4°C, there is a tolerance range of 0°C to 6°C. Obviously, freezing a sample will minimize further the microbe activity, but it might also cause all your sample containers to break.

When the lab receives the cooler, they check the temperature in a couple of ways. Some may quickly place a thermometer in the cooler while they sign the COC. That temperature is recorded on the COC. They may use another thermometer that checks the surface temperature of a sample container. Or, there is an infrared thermometer that they point at a sample container that provides the temperature of the sample. In all methods, the temperature is recorded on the COC and often provided in the analytical data. If the temperature is greater than 6°C, the sample can be called invalid, and back to the field you go—at your expense.

There are several considerations regarding achieving and maintaining the prescribed temperature. The most common method used to cool samples is ice, but there are several precautions with this. First, you should buy ice in the morning of your sampling event and put it in the coolers in order to chill the bottles before filling them. If you wait until the end of the day before buying ice, even if the samples are in the cooler, they will not be at 4°C, thereby nullifying your efforts. Granted, the lab may not know this when they receive the samples and record the temperature, but this is not acceptable.

I use ice cubes to cool coolers. They are available for purchase at convenience stores, grocery stores, hotels, some fast-food restaurants, and various other places. However, you do not just pour the ice into the cooler, you have to put the ice into sealable (e.g., Ziplock®) bags. I usually double bag them. There are several reasons for putting ice into plastic bags. First, ice melts at temperatures above freezing. If the melt water is uncontained in the cooler, it can cause labels to peel away from their sample containers and/or can cause ink to run off of the labels. In either case, the samples will be invalid because you will not know from which container the label peeled, and you will not have the sampling time and location on the labels.

Another issue is that melt water can leak out of the cooler, causing concern with the transportation company (e.g., Fed-Ex). I have heard of samples sitting at Fed-Ex for several days because the cooler leaked and the workers did not want to handle the cooler. They had no idea if it was ice water, contaminated water, or some other substance. Putting ice in sealed bags will prevent these scenarios from happening. In transportation, a bag filled with ice water might burst open, so I double bag them to prevent this.

The following was found at the Fed-Ex web site:

a. FedEx does not recommend the use of wet ice (frozen water) as a refrigerant. If you believe wet ice is necessary, please call the FedEx Packaging Design and Development Department at 800.633.7019 for specific packaging requirements. Use of wet ice without preauthorization is prohibited.

b. If a shipment is refused by the recipient, leaks, or is damaged due to inadequate packaging, the shipment will be returned to the shipper, if possible. If the shipper refuses to accept the returned shipment or it cannot be returned because of leakage or damage due to faulty packaging, the shipper is responsible for and will reimburse FedEx for all costs and fees of any type incurred in connection with the storage or disposal of the shipment or the cleanup of any spill or leakage from the shipment.

Another consideration with preserving samples by temperature is the use of an insulating/protective sleeve for the sample container. VOC vials are supplied like this. The sleeve may be bubble wrap, foam, or some other material. These are good to use for sample protection during transportation, but you should place the sample on ice before putting on the sleeve. If you collect a groundwater sample with a temperature of 60°F and put an insulating sleeve around it, you are insulating the sample at 60°F and it may not fall to the prescribed 4°C. So, put it on ice then, and when getting ready to ship it, put on the protective sleeve.

Acids and Bases

The typical acids used for sample preservation are hydrochloric (HCl), nitric (HNO_3), and sulfuric (H_2SO_4). Hydrochloric acid is used primarily for VOC preservation because it "traps" the target contaminants, keeping them from changing. In addition, HCl prevents biological activity.

Nitric acid is used for metal preservation and it keeps the metals in solution by lowering the pH. Without the acid, the metals could precipitate, thereby affecting sample results.

Sulfuric acid is used for several water quality parameters and it serves to minimize biological activity.

The only base used is sodium hydroxide (NaOH) and it is used for cyanide analysis. If CN is in an acidic environment, it can be released as gas, therefore, it is maintained at a pH >12. When the lab receives the sample, in order to release or extract the cyanide, they add acid to decrease the pH.

Some labs provide containers already preserved. If your lab typically does this, double check the containers and add labels before you go to the field to assure that the right preservative is in the right container (as shown in Table 7.1). Some labs provide an ampule of preservative for each container, which you have to add yourself. In these cases, I have typically had better success adding the preservative to the containers before I get to the field. This is especially true when sampling in inclement weather. It is very difficult to snap open the glass ampule when your hands are freezing or in pouring rain. Also, a strong wind can redirect the preservative onto your skin, clothes, and eyes when pouring it into a container. So, a controlled environment is the better place to add preservative.

When going on large sampling projects that last for several weeks, I have received preservatives in bulk (250 - 500 ml) that I had to add to the containers. This will usually come with several eye droppers and instructions as to the volume to add. Again, do this in a controlled environment if at all possible: spilling a large volume of acid is not good. The negative side of receiving bulk preservative is that you are usually provided with an abundance of material. Upon completion of the project, some preservative may remain, so check with the lab to find out if you can send it back to them. Otherwise, you might have to dispose of some preservative that is characteristically hazardous, due to its pH being less than 2 or greater than 12.5.

On some projects, like Superfund, you might be required to add preservative and/or check the pH after collecting the sample and prior to submitting it to the lab. Your project manager and federal contact will inform you of this requirement, which may vary by region. Find a controlled environment to do this, like at the end of the day in a staging area. If the pH is not low enough, additional preservative will have to be added. Better to do this in a warm, dry staging area than in the middle of a wind swept plain. When checking the pH, to prevent cross contamination, do not place the litmus paper into the sample. Either carefully pour the sample onto the paper or use an eye dropper or a clean, stainless steel spoon to take the liquid out of the container and place onto the litmus paper. You will have to decontaminate the eye dropper, spoon, or whatever else you used between checking sample pH. Never check the pH of a sample collected for VOCs (VOA Vial). Once that sample is closed and no bubbles are present, keep it closed.

Sample Containers

Specific sample containers are chosen so that they preserve or do not affect a sample. Several analytes require amber containers because they photo-react, meaning that light affects them. Some analytes can absorb into plastic so they have to be placed into glass sample containers. The number of containers and sizes required are based upon the volume needed to conduct an analysis. For example, SVOC water analysis requires one liter of sample. You are often required to collect two liters so additional testing can be conducted, if necessary. VOCs are collected in 40-ml vials because you do not need much volume for analysis and it is more difficult to get zero headspace in a larger volume. You typically collect two vials to allow for the possibility of sample breakage during shipping.

Holding Times

These are determined based upon the length of time you can keep a sample prior to analyisis and be assured that it still represents conditions or characteristics at time of sampling. Some holding times are immediate, such as with chlorine or sulfite, and some are within 24 hours (hexavalent chromium). VOCs have holding times of days while most metals are months. Also, holding times can vary based upon the sample matrix (soil or water). For TCLP analysis, there are two times provided: holding time to extraction, and holding time to analysis. Depending on the parameter, labs have between 14 and 180 days to perform the extraction. Extraction means that if a lab is doing a TCLP benzene analysis, they have to remove or extract the benzene from the soil. Upon extraction, the lab has between 14 and 180 days to analyze the sample regarding the concentration of the benzene extracted.

You only have a certain amount of time before a sample has to be submitted to the lab, before the lab has to prepare the media for analysis, and before they have to analyze it. The sampler's role is to get the samples to the lab on time. Generally, you ship the samples priority overnight at the end of each sampling day so they arrive at the lab early in the morning. The lab will then put the samples in a refrigerator to maintain their temperature at 4°C.

For select analytes, hexavalent chromium and tritium, the holding time is 24 hours. You should sample for these analytes in the afternoon, just before you ship them to the lab. If you sample for them at 8:00 AM, they may not get to the lab until 10:00 AM the next day, thereby exceeding the holding time.

Quality Assurance

The overall quality assurance (QA) objective of a sampling event is to develop and implement procedures for field sampling, laboratory analysis, chain-of-custody, and reporting that will provide results that are legally defensible in a court of law. Certain quality assurance samples are collected to demonstrate precision and accuracy. Precision demonstrates the ability to reproduce the measurement process. Accuracy is the measure of agreement between a sample value and its accepted reference value. Precision and accuracy in the field are measured through duplicate samples, equipment blanks, trip blanks, and field blanks. They can also be measured through regular and appropriate field instrument calibration and adherence to sample holding times. The precision and accuracy of laboratories is measured through MS/MSDs and mathematical analysis of the lab data.

MS/MSD

The matrix spike/matrix spike duplicate (MS/MSD) is collected to assess the lab's ability, specifically its accuracy and precision, to analyze a sample. In the field, you must fill additional sample containers for this analysis to be conducted. To explain this further, consider an SVOC groundwater analysis. You usually have to fill a one-liter amber sample container for SVOCs. If you are required to collect for an MS/MSD, you would have to collect two additional one-liter containers. The MS/MSD should be collected from an area where contamination is expected.

When the lab conducts the analysis, they will fortify (spike) the matrix spike with a known quantity of SVOCs. The spiking compound is dictated by the SW-846 method, which provides a list of spiking compounds when conducting SVOC analyses. For discussion purposes only, say the matrix spike was spiked with 10 ppm of pyridine. That sample is analyzed to determine if pyridine was detected. This test is conducted because there might be something in your sample that causes interference during analysis that would prevent pyridine from being detected. If pyridine was not detected, the chemists in the lab would then have to adjust their analytical method to ensure that pyridine could be detected.

With the matrix spike duplicate, the lab again spikes the sample with SVOC compounds at certain concentrations. Knowing that their method can detect pyridine, they will then assess the precision of the method. If they spike the sample with 10 ppm pyridine, they would expect 10 ppm pyridine in their results. The lab is allowed a range on these results. (See the section on Data Validation below.)

Surrogates

While this is not a sample that is collected in the field, this topic blends nicely into the MS/MSD discussion. Labs not only fortify (spike) samples for MS/MSD purposes, they also fortify samples with surrogate compounds and during extraction and analysis, and they assess their ability to extract these surrogates from the sample. Many compounds for which you analyze have specific surrogates that closely resemble the actual compound. The idea is if the surrogate, which is injected at a known concentration, is not extracted sufficiently, then the target compound might not be, either. The surrogate will alert the lab chemist to possible masking or interferences.

There are specific surrogates for each type of analysis (VOCs, SVOCs, etc). The difference between a matrix spike and surrogate spike is that MS/MSDs are spiked with compounds for which you are analyzing while surrogates are not. For example, in VOC analysis (SW-846 Method 8260b), matrix spikes must include 1,1-dichloroethene, trichloroethene, chlorobenzene, toluene, and benzene, which are constituents on a typical target compound list. In addition, the matrix spiking solution should contain compounds that are expected to be found in the types of samples to be analyzed. For example, if you're sampling at a gas station, the matrix spikes might include benzene and toluene, both of which you would expect to find in samples from a gas station. Surrogates for Method 8260b include toluene-d, 8, 4-bromofluorobenzene, 1,2-dichloroethane-d, and dibromofluoromethane. These are not compounds you would expect to find at a site. Nor are they samples you would find on a target compound list. They merely imitate the behavior of the target compounds.

Duplicate

In the field, two samples are collected from the same location at the same time. One is labeled accurately to reflect the sample location and the other is labeled with a fake sample location. For example, if you are at MW-1, you begin by collecting the VOCs. Then, you collect the duplicate VOCs. You proceed to collect the SVOCs then the duplicate SVOCs. You continue until all the parameters have been collected. Then, you label the original sample as MW-1 and the duplicate with a fictitious identifier. For example, if your site has eight monitoring wells, label the duplicate as MW-9. Keep in mind that you are not likely to fool the lab, in that the samples will be labeled with the same or similar times. Regardless, the purpose of this is to see if the lab can replicate the first sample results in the second sample results. Again, they are allowed to be within a certain range to demonstrate precision.

Field Blank

The field blank is collected to assess field conditions that may affect your sample. It is collected by filling sample containers with distilled or de-ionized water while you are at a sample location. These samples tell you if the conditions in which you are sampling are adversely affecting your samples. For example, if you are using a generator to operate equipment and are sampling within the exhaust plume, that plume may add contaminants to your sample. Likewise if you are in the plume of an industrial smoke stack or if you are sampling in a busy downtown area with a lot or cars, buses, and truck exhaust around you. If you are sampling at a gas station and people are filling their cars with gas throughout the sampling event, the gas vapors may affect your sample. Also, if you are wearing insect repellent, perfume, or cologne, it may get into the sample so field blank can assess those anomalies, as well. So, if you get a hit of benzene in all of your samples and in the field blank as well, it may suggest that the sample concentrations are a result of influences associated with the ambient conditions.

Equipment Blank

This sample is collected using the equipment you use to collect your samples. Its purpose is to assess your ability to decontaminate the equipment. The procedure is, after collecting a sample in a contaminated area, you decontaminate the equipment that you used. If you used a bailer, you would fill the bailer with distilled or de-ionized water and then fill the sample containers with that water. If you used a pump, you would have to put the pump in sufficient distilled or de-ionized water, then fill the sample containers. I have found that a five-gallon water cooler jug of distilled water will accommodate most pumps, although you may have to cut a larger opening in the top. If you did not do a thorough job of decontamination, there may be contaminants in the field blank.

You should run this sample after you collect a contaminated sample. It is a waste of time and money if you collect it after you get a clean sample. Also, if you use dedicated samplers or disposable samplers, neither of which requires decontamination, then you do not have to collect an equipment blank.

Trip Blank

The trip blank is a set of VOC vials containing distilled or de-ionized water and is analyzed for VOCs. Typically, the lab prepares this set and sends them in the sample coolers often taped to the lid. If you do not have a set and need one, you can prepare your own, ideally before you go into the field, by filling a set of VOA vials with distilled or de-ionized water. You keep the trip blank set in the cooler with all the other VOC samples. The lab will analyze the trip blank to assess if any contaminants have affected the cooler during transport. For example, if the cooler was left on a shipping dock for hours, the exhaust may penetrate the cooler and samples. The trip blank would reflect this penetration and cause the remaining sample analyses to be suspect.

Chain-of-Custody

The Chain-of-Custody form is quite possibly the most important form used during sampling. It lists all the samples, the sample matrix (e.g., soil, water), the parameters for which they are being analyzed, the dates and times sampled, specific instructions, company information, cooler temperature, and signatures of those persons who collected and handled the sample. As suggested by the name, the primary function of the Chain-of-Custody form is to list the people who had possession (custody) of the samples and when they had them. When you complete the Chain-of-Custody form and are ready to give the samples (relinquish custody) to anyone else (a co-worker, the lab), you have to sign (with date and time) that you are no longer in custody. Then, the person who takes possession of the samples (gains custody) has to sign, with date and time, thereby indicating that they have received them. The idea is to demonstrate that the samples have always been in somebody's control. If someone relinquishes custody at 7:00 AM on a given day and the person who next takes the samples signs at 10:00 AM, that indicates that nobody had control of the samples for three hours. This is enough to invalidate the samples, because no one can verify that the 4°C temperature was maintained or that no one tampered with the samples (e.g., replaced a contaminated sample with distilled water). Among the first things scrutinized in environmental litigation is the Chain-of-Custody. If it is not complete, not completed properly, or not present, then the entire sampling event is in question.

Custody Seals

Custody seals are placed on sample containers and/or coolers to demonstrate the integrity of what has been sealed. You are required to sign and date each custody seal. The seal is a strip of paper with an adhesive back that is placed on the cooler or sample so that if anyone tries to open the sample, the seal breaks, suggesting that the samples may have been tampered with. The seals are designed so that they cannot be put back together again if you break one or try to peel one away.

Standard projects might only require custody seals on the transport cooler, while other projects require custody seals on each sample container. The exception is soil samples collected using Method 5035. The sample vials are pre-weighed, so adding a custody seal would adversely affect the analysis.

Data Validation

As I stated in the beginning of this chapter, your responsibility for valid data is relinquished briefly when the lab receives the samples and conducts the analyses, but returns to you once analysis is complete and the data are delivered to you. That is not to say that valid data is not a concern to the lab, but at that point it is out of your hands. The lab has many criteria and protocols to which they must adhere to provide valid results. When they have completed their job and delivered the analytical results, it is then up to you to check the data to ensure that the analyses were conducted properly. The process of checking lab data is called data validation.

The level of effort for data validation could range from several minutes to several weeks, depending upon the requirements and size of the project on which you are working. At a minimum, when you receive the results, you should check that the samples were analyzed within the designated holding times. You could also check that the correct detection limits were used. For example, if the regulations you follow require a detection limit of 10 µg/kg for a certain compound and the lab has reported the result as non-detect at 50 µg/kg, then there is an issue.

Beyond those two examples, there is not much data validation you can do without a data package. A data package is raw data that the lab generates during analysis and is used to determine the concentrations of your sample results. There are varying levels, with different amounts of information in each. Perhaps the most common data package terms are Levels I, II, III, or IV, with Level IV being the most complete package. The Level IV package would include all raw data including chromatograms, mass spectra, equipment calibration records, quantitation reports, reconstructed ion chromatograms, and data system printouts. To get a better idea of what that is all about, go to *http://www.epa.gov/superfund/programs/clp/guidance.htm*, which provides links to data validation guidance documents.

A full data package (e.g., Level IV) is typically required on the most complex and litigious projects, while more routine projects might not require any sort of data package. Also, a full data package is needed to support a site risk assessment. Of course, there are packages for those projects in between. The need for a data package is usually determined long before you go to the field to collect samples, and requires considerable time to complete after the samples are analyzed. Labs charge extra for the data package with a complete package cost being about equal to the analytical costs.

A complete package for a large project will comprise hundreds, if not thousands, of sheets of paper. To review those data is a huge undertaking and could take days to complete. It is not necessarily difficult, but it requires attention to detail, good note taking, patience, organization, and a large workspace. It also requires professional judgment and a good understanding of laboratory analyses, which develop with time. Until that time, you can hire firms that have special expertise in performing data validation.

The data validators must look at the paperwork associated with each sample collected, including the blanks, duplicates, and MS/MSDs. They must also look into the paperwork for the instruments that the lab used in the analyses to assure that they were calibrated properly and in a timely manner. If the analyses were conducted over several weeks, there would be several weeks of calibration records. Also, the lab runs equipment, method blanks, and control samples to ensure that the instruments are decontaminated and working properly. Those records require review as well.

The reviewer is looking for various things. With an MS/MSD sample, a reviewer is looking at the recovery of the spiked compounds. Each compound has a range or limits, based on percent, that has to be "recovered" during analysis. Likewise with surrogates, each compound should be recovered within a certain range. With blanks, the reviewer looks for compounds detected in these samples, which should have none. For equipment tuning/calibration records, the reviewer looks at "ion abundance criteria" of the calibration standard (e.g., bromofluorobenzene) to assure the equipment is working satisfactorily. During the data review, there are ranges to check, statistical analyses to run or check, equations to solve, and notes to record.

If the reviewer finds problems with the data, the associated sample(s) is "flagged" to denote the problem. The flag is actually a letter next to the analytical result that represents the issue. Perhaps the most common flags are the "U" and "J." The "U" does not denote a problem, it simply indicates that the analyte was not detected. The "J" indicates that the analyte is present but the value is estimated. "J" is often found with samples with high contamination that have to be repeatedly diluted to obtain the value. Conversely, if an analyte is detected at a concentration less than the detection limit, which is possible, that result will be flagged with a "J" to denote that it's estimated. The standard "flags" from EPA guidance documents are as follows:

U - The analyte was analyzed for, but was not detected above the reported sample detection limit.

J - The analyte was positively identified; the associated numerical value is the approximate concentration of the analyte in the sample.

N - The analysis indicates the presence of an analyte for which there is presumptive evidence to make a "tentative identification". This is for tentatively identified compounds (TICs).

NJ -The analysis indicates the presence of an analyte that has been "tentatively identified" and the associated numerical value represents its approximate concentration.

UJ -The analyte was not detected above the reported sample detection limit. However, the reported detection limit is approximate and may or may not represent the actual limit of detection necessary to accurately and precisely measure the analyte in the sample.

R - The sample results are rejected due to serious deficiencies in the ability to analyze the sample and meet quality control criteria. The presence or absence of the analyte cannot be verified.

Throughout this section I have indicated that data validation is conducted back in the office upon receipt of the analytical results and data package. Please note, however, that laboratories do conduct their own data validation and your sample results, even on the "low profile" projects, might come already flagged. The lab is not going to wait weeks or months for someone else to check their work.

8

The Ideal Tool Kit

You always need the right tool for the job, and each job may require many and varied tools. I have wasted too many hours using a screwdriver as a hammer, a hammer as a pry bar, and a pry bar as a screwdriver. The solution is somewhat obvious, but escapes so many people. You need a good tool kit...an ideal tool kit. The ideal tool kit has in it most things you will need on a job.

This concept was introduced to me on a three-month long geotechnical drilling project that went 24 hours a day, seven days a week. During the preparation time prior to drilling, we went to the local lumber yard and built boxes for tools. There were four drill rigs, so we built four boxes that had hinged lids and were lockable. We then went to a hardware store and bought four of everything we needed, including duct tape, hammers, screwdrivers, magic markers (for labeling core boxes and Shelby tubes), electrical tape, and putty knives. We added hydrochloric acid to test for the presence of limestone/dolomite. These boxes stayed with their respective drill rigs at all times. When the shift changed, the next person knew he had all the necessary tools. The project manager of the job called these toolboxes "happiness kits," because if you did not have them you would not be happy.

The Container

I do not recommend a wooden box for your tools. Many samplers have small vehicles, and a large wooden box would be cumbersome and may even create back problems when attempting to lift it from your trunk. For people with sedans, wagons, SUVs, or pick-up trucks with caps, I recommend a sturdy plastic container (e.g., Rubbermaid®). You can get these containers at most big retail chains or The Container Store®, as they are noted for selling containers. These containers can be stored securely in a locked vehicle, which is important. If you drive a pick-up truck without a cap, I recommend a metal toolbox designed for the truck because, if you go with the plastic container, you would have to move it into the cab every time you stop for coffee, lunch, etc. If you have a metal toolbox, I recommend partitioning your tools with smaller plastic containers.

There are various ways to organize your tools in the large plastic container. I recommend supplementing with smaller plastic containers which make finding your tools an easier task. You could keep all your screwdrivers in one container, or Philip's in one and standard in another. You could keep miscellaneous ratchets in a separate container. You could have a small container for miscellaneous screws, nuts, and bolts, another for various pens, pencils, and markers. This method is unnecessary if you buy a 100-200 piece tool kit that is self-contained. However, as you add various pieces, the separate container method is beneficial.

If you choose the separate container method, clearly label each container with its contents. Either write on top of the lid, or put duct tape on the lid and mark it. Try to stay loyal to the same brand of smaller containers; they are usually designed to stack neatly on top of each other. This sort of detail will ultimately save money and hours in the field. I have searched high and low for a tool, been unable to find it, then gone to a store to buy a replacement. Inevitably, I would find it later in the day. It is frustrating, unproductive, and expensive and I cannot justify charging duplicate tools to the company because I have misplaced the original in my car.

The Contents

For the typical environmental project, the tool needs are different.

- A scale—I prefer a thin, bookmark style scale for the field. I find mine at my local survey supply firm. The one I use is usually a free item located next to the cash register. Another option is a miniature triangle. This, too, I ususally acquire for free, as it advertises a company on its side. You can use a standard, foot long triangular scale but the others are more portable and convenient.

- A geotechnical gauge (as mentioned in Chapter 1)—This can be purchased from Forestry Supply (*www.forestry-suppliers.com*) and costs about $15. It is a great tool to have when soil sampling. It has a color guide, particle size guide, particle shape guide, rulers, and other soil classification tips.

- A Fisher Space Pen®—This pen was developed for NASA to be used on the Apollo missions. It is pressurized, so it can write at any angle, including upside down. It writes in the rain, snow, and freezing temperatures.

- A speed wrench—This is used to tighten drum lids and requires a $^{15}/_{16}$ ratchet. I have used channel locks and vice grips to tighten and loosen drum lids, but they take too long and are frustrating. Spend the money on the speed wrench, and save the hassle of the alternatives. I actually once asked a project manager if I could buy one and he said no. If such is the case with you, buy your own—it will be worth it.

- Duct Tape—No explanation necessary.
- Allen wrenches—You will eventually have some equipment that requires Allen wrenches. Get Imperial and metric, you will need both.
- Carpenter's Rule—This is a wooden folding ruler. It is great to put on the table where you are logging soil samples. It provides a quick measurement to get the depths of various strata. These rulers come with measurements in tenth's of a foot or in inches. If you cannot find one with both measurements on either side, buy two. The tenths are great for borehole logging; borehole logs are measured in tenths, not inches. These also provide a sturdy measure, as opposed to floppy steel tapes.
- Tape measure—Get a 25 foot floppy steel tape. They are good for measuring down boreholes and other, more obvious applications.
- 100-foot tape—Get the fabric type that winds up. The types with fluorescent orange cases are easier to find at the bottom of the Ideal Tool Kit.
- 200 or 300-foot tape—They are better for setting up 100-foot grids. They are also easier and more accurate than 100-foot tapes for longer measurements.
- Screwdrivers—Get a variety of sizes and types; standard, Philip's, star, and Robertson's. Buy a brand name that can be replaced free of charge. Do not use them as hammers or pry bars or you will have to get them replaced.
- Pry-bar—A two-foot long pry-bar is very handy for lifting manhole covers, monitoring well caps, and steel plates. A four-foot long pry bar will move more, easier, so get one if you have room to store it.
- Hammer—Buy a good hammer. I love Estwing® hammers. They are steel, durable, and feel great in your hand. They come in several sizes and have various purposes. (There is even an Estwing geology pick, if you decide this is your favorite brand as well.)
- Wire strippers—When you are an environmental sampler, you are also an electrician. You will need these to repair pumps, engines, and other electrical equipment.
- Bolt cutters—Also known as a universal key. You will forget keys and you will go to new sites with other people's locks. These will get you in. I have never used bolt cutters to cut a bolt, only locks.
- Lock de-icer—For those of you in colder climates, you will need this at some point.

- Propane torch (portable)—For when you forget the lock de-icer. Also good to melt ice from flush-mounted monitoring wells.
- Wrenches—Always essential.
- Pliers—You will need various shapes and sizes (needle nose, flat nose, etc.)
- Channel locks—Get various sizes.
- Vise-Grips—Get various sizes. These are as useful as duct tape.
- Utility knife—Good for cutting hoses and rope. Keep spare blades handy and change the blade often.
- Hack saw—There will inevitably be metal to cut.
- Pipe wrenches—Get several sizes. They are needed for holding onto pipe when constructing a monitoring well. Also, they are necessary when loosening or tightening pipe (PVC or steel).
- Leatherman®—This is a beneficial tool that you can keep on your belt. It has pliers, screwdrivers, wire cutters, a knife, and other features. It is good for when you are unable carry the entire Ideal Tool Kit into a field. I have no brand preference; others make quality utility tools (Gerber®, Craftsman®, etc.).
- Teflon® paste and/or Teflon tape—Sampling will often include putting pipe together. Make sure it does not leak and that you can get it apart again. Teflon will assist in that purpose.
- Electrical tape—For wire repairs. Avoid putting any type of adhesive tape down a monitoring well. There is a possibility of contaminating the water. If you need to repair wires that will go into a monitoring well, use heat shrink (available in the electrical section) to protect the wire from water.
- Flashlight—I like the long black flashlights (Mag-Lite®). They have lots of power and are sturdy.
- Mirror—Buy a compact or small mirror. These are great for reflecting sunlight down wells, manholes, and boreholes. They can only be used on sunny days but you will be surprised by the supplied luminosity.
- Hand cleaner—Put in an abrasive waterless hand cleaner. They are often citrus-scented and are great for clean-up after engine repairs, digging, etc.
- Gloves—Include Nitrile® and leather gloves.

- Clipboard—Except when logging boreholes, I use a metal clipboard that opens. They hold a lot and are durable. I have had mine for 12 years. For borehole logs, use the clipboard described in Chapter 1, Geology Fundamentals.
- Paper towels—Needed for cleaning hands, cleaning tools, wiping brows, and wiping tape measures.

The Guidelines

Where applicable, buy the above tools that may be returned if broken (e.g., screwdrivers, wrenches, etc.). You should do this for two reasons. First, companies that make tools that can be returned if broken seem to make the tools very well, so that they do not break very often. Second, when they do break, you can return them. Too many stores sell a variety of tool packages where you can buy 99 pieces for a ridiculously low price. Based upon my experience, these tools tend to bend, rust, and break easily. When you are 20 miles from the nearest town and you need to fix some equipment, it is much better to have tools that do not exhibit these properties. Returnable tools include Craftsman from Sears®, Husky® from The Home Depot®, and Snap-On® tools.

I do recommend packaged tool sets with 99 tools in a plastic container when they are made by the brands listed above so the broken pieces can be replaced free of charge. They are a handy way to store 99 tools and avoid using a lot of smaller containers. They are also a less expensive way to start a tool kit. You will usually find them on sale just before Father's Day and Christmas. There is a certain mindset that comes with the tools you buy. If you have the low cost set, you may treat them as such. In your mind, they are disposable, so if one falls in the water you may not be as apt to dry it. You are also more apt to lend out your tools and forget about them. If you buy an expensive set, you will take care of them and use them for their intended purpose, which will make them last even longer.

The Field Book

This is a crucial document that can be used in a court of law. The field books I used are the orange survey books that measure approximately 7-by-5 inches. The pages are bound (often sewn) such that you cannot tear them out without it being evident. I buy hard-back books simply because my writing is neater in those books. I prefer to use books with pre-numbered pages so I can demonstrate clearly that the pages have not been torn out. This is important for litigation cases where you and your field book are on the witness stand. Typically, you can buy these books from a survey supply store (e.g., *www.kara.com*). EPA guidelines recommend not only that field books are pre-numbered, but also that there are lines for your signature and the date on each page. This is to justify further the field record being presentable in court.

In addition to page numbering and not tearing out pages, there are numerous rules that one should (and often must) follow when using a field book. The rules are in place for the time that your field book will be cross-examined. The more you write in a field book, the better off you are when you get back to the office. Questions will be asked of your day in the field, and if you are unable to answer, you are supposed to review your notes. A Quality Assurance Project Plan (QAPP) guidance document from the EPA provides these tips, and others come from experience.

1. Sign each page.
2. Record the date on each page.
3. Record the weather as you begin the day. As it changes, record that as well.
4. Record times you arrive and leave site.
5. Write in pen (pencil can be erased).
6. Cross out mistakes with a single line and initial the mistake. Do not scribble over your mistakes because the opposing counsel might think you are hiding something.
7. Record the times that you sample. When you get busy, you may forget to write the time on a sample container. Refer to your field book to get the time.
8. Record conversations you had with the project manager or client. This is especially crucial when direction is given. If the manager or client tells you not to analyze for metals in sample 4, this should be noted with the date and time. It is better if you can record why the protocol was changed, too. Two weeks later, when the results come in, somebody will be looking for the metals results for sample 4 and you will have to explain why they are not there.
9. Record any changes from the original Scope of Work, Workplan, or Field Sampling Plan (FSP). If you are given a precise location to sample and a car is parked there, note that, and note precisely to where the sample location was moved. However, before you relocate the sample location, check with your project manager. You may arbitrarily move it to the south and he may want it to the north.
10. Record that you calibrated the various instruments that you used.

11. Record the people on site, including your co-workers, oversight contractors, sub-contractors, and visitors. Note the time they arrive on site and the time they leave. Record pertinent conversations that you have with them.

12. Log daily activities. If a piece of equipment breaks down, record the time it broke and when it was repaired. Later you can make sure you are not charged for it's time during breakdown. If you stop activities for any reason, write when, why, and for how long. On these projects, there are often billing discrepancies. The more detail you have, the easier it will be to justify billing.

13. Whatever you write, keep it factual so that it will be presentable in court. Many people try to be funny, sarcastic, poignant, etc., but the field book is not the place for this. Never write disparaging comments about your co-workers.

> Career Tip—If you can maintain accurate logs for billing purposes, you will get bigger jobs to manage and more responsibility, ultimately leading to more money.

> Career Tip—He who keeps neat, clean, concise, and accurate notes will go far. Your notes, like it or not, are a reflection of you.

There are a couple of perspectives on field book use. Some companies will have a single field book for a specific job while others will assign a field book to an individual for various projects. With the first perspective, each book used on a project is numbered consecutively, and the start and finish dates are marked on the front of the book and the spine. As the books are filled, or when they are not in use, they are stored in the project file. This method is beneficial for when you need to look back at your notes or somebody else's notes. They are available for reference. With projects using surveying, the benchmark data may be in a previous book, but they can always be found and referenced, as required. If there are several crews on a large project, each crew lead will have a numbered field book to record data.

The other method employed is each person having their own field book. This is good for smaller jobs and for the person who may work on many projects. These jobs have the same legal requirements as a large job, but using a single book for three or five pages is impractical. So, for this book, your name and the start/finish dates should be marked on the cover and spine. It is imperative that the relevant project number or name be recorded on each page so the information can be easily identified.

The Items that Do Not Fit in the Ideal Tool Kit

A few of the measuring devices that you have with you at all times are handy tools to have. They are your hands, your feet, and your paces. With your hand spread out, the distance from the tip of your thumb to the tip of your little finger is unchanging (barring any unforeseen accidents). On my hand, for instance, the distance is nine inches. When you need approximate measurements, this is a fine tool to use. When you forget the measuring tape or do not want to run back to the car to get it, your hand span may suffice.

My foot, with a work boot, is one foot long. When measuring short distances (five to ten feet), I will walk heel to toe, heel to toe to get the measurement. Again, it is approximate, but will suffice in many cases. It is convenient for measuring floor tiles. It is widely accepted that floor tiles measuring seven or nine inches square are asbestos-containing material. With my foot, when I first walk into a room, I can assess the presence of tiles with relation to my "foot."

Finally, the last appendages I use for measurement are my legs. When walking a normal stride, my pacing factor is 5 ½ feet. There are two ways to measure your pace. I use "left right left." Starting with the toe of my left foot, I walk forward with my right then left foot again, the distance from left toe to left toe is 5 ½ feet. Others use one step as their pace (left then right and measure). Either is fine. I use my pacing factor when measuring longer distances of five to several hundred feet. It is ideal for approximate property boundaries, distances from boring to boring or to sample locations, etc. For me, nine paces measures approximately 50 feet, and 18 paces measure about 100 feet. These bodily measurements are estimated measurements. For precision, break out the measuring tape or the survey crew.

To get your pacing factor, lay out a 100-foot tape. Walk along the tape using your normal stride and count the number of paces or steps, making sure to not overextend. Repeat this five to 10 times. It may vary slightly each time, so keep doing until you get your average pace. Once you have a good average, remember it or write it down. When determining and using your pacing factor, use your normal stride because you may have to walk through heavy woods/underbrush to get a distance. One "giant step" would be difficult in these circumstances. When I worked in Canada, I used a metric pacing factor. In the US, I use Imperial measurements.

The Cost

You may not be able to run out and buy everything in the Ideal Tool Kit, but be patient and develop your kit over time. If you buy all the tools yourself, you can bring it with you if you leave your company. With some companies, you might be able to charge your kit to the job to help recover your costs (e.g., $10/day). It is a fuzzy line, deciding what the client, the company, and the sampler should pay. Site-specific tools that may only be used on one project should be purchased by the client, but leave them with the client upon completion. If you think you will use a tool again, charge it to your company or buy it yourself and take it with you. Or, have the company buy it, but store it in their "tool area" for common use.

It is impractical for a company to hire someone and expect them to have all the tools of the trade. They should supply a tool kit for common use, just as they supply computers, desks, pens, etc. However, the nature of the business is that someone will use the kit, break tools, and not tell anyone or replace them. When you grab a kit to go in the field, you will be surprised and incensed when all the standard screwdrivers have their heads twisted off. It is very embarrassing when your boss or client is watching you fumble for a simple standard screwdriver. When you get back to the office, nobody will admit to breaking or abusing any tools.

For this reason, I recommend you start your own Ideal Tool Kit and keep it with you. You can develop your own policies for lending your tools. Write down to whom and when you loan a tool. Have them return it to you, not to the kit so you can keep track of each tool.

Career Tip—You will do a better job if you have and can rely on your own tools. People will want you on their projects because you will be more efficient.

Other Supplies

There are many other items that you may need for your sampling efforts, above and beyond the tool kit. These supplies are just as crucial in some cases, but may vary from job to job. You should have the Ideal Tool Kit on all jobs. Other items include decontamination equipment, sampling equipment, and PPE.

Decontamination Equipment

1. Long and short handled brushes - These are used to clean various types of equipment including sampling spoons, bowls, knives, etc.

2. Bottle brushes—These are used to clean inside bailers, assuming you are using stainless steel bailers. If you are using disposable bottle brushes, throw them out, do not decontaminate.

3. Plastic sheeting—Useful to contain your liquids and maintain your decontamination area.

4. Paper towels—I have worked with some people who have preferences in paper towel selection. I do recommend you buy a national brand that does not leave much lint or dissolve in water. If you buy a cheap brand, you will spend more time going out to buy more when they all dissolve in your hands.

5. Plastic tubs or buckets—I have had success with five-gallon buckets to decontaminate equipment. I have also had success with black plastic tubs, available at most hardware or automotive stores. The larger tub is good for soaking a lot of equipment. Note that it's best to avoid galvanized equipment so as to prevent metal cross-contamination.

6. Pressurized sprayers (H_2O)—I do not like these. They are steel sprayers that you hand pump to pressurize. I find that since it has several working parts, there are several parts that do not work when they need to and they are cumbersome. When left in vehicles overnight in winter, they freeze. And, they take forever to thaw in the front seat of your car with the heat on high. The pump handles bend, the nozzles become blocked with dirt or ice, and the lids become stuck. Besides, the amount of pressure they produce is not much greater than a spray bottle and it does not last very long.

7. Solvent sprayers—These, I like. They can be bought in most hardware stores and are usually found in the garden section of your hardware store. My rule for these spray bottles is to rinse them thoroughly with water before using to get rid of any residual contaminants that may be present from the garden center.

8. Aluminum foil—Used to wrap equipment upon decontaminating.

9. Non-Phosphate soap—The industry standard seems to be Alconox®. You go to *www.Alconox.com* to obtain dealer locations near you, or buy it online from them. There are several detergents besides Alconox. The one with which I am most familiar is Liqui-Nox®, which is a liquid-based detergent, as opposed to the powder form of Alconox. Personally, I prefer Liqui-Nox, because it appears to dissolve easier in water, especially on colder days.

10. Solvents—The type of decontamination you do may depend on the analysis you will be conducting on the samples collected. I usually keep distilled water in my car. I do not travel with other decontamination solvents (e.g., methanol, nitric acid, acetone, etc.) unless I am going on a job.

Waste Disposal

1. Trash bags—I have seen many varied uses for trash bags on sampling events. First, they are used to hold trash. Second, they are used as plastic sheeting when you run out of Visqueen®. Third, they are used as raincoats. Finally, they are used to wrap grossly contaminated equipment so it does not get all over everything or used to wrap clean equipment so everything does not get all over it. For clean equipment, I have had

people tell me not to wrap it in garbage bags for fear of contamination from the bag. This may be accurate, so be cautious in using a trash bag for that purpose. Keep in mind the analytes for which you are sampling, however. If you are sampling for BTEX, contamination from a garbage bag is not likely to affect your results. If you are sampling for the full suite of SVOCs, perhaps it is more of a concern. If you are sampling at a garbage bag manufacturer and trying to demonstrate that they are impacting the environment with plastic resin, do not use the garbage bag.

2. Trash containers—These are good for holding trash or a lot of water. They can also be used to decontaminate large pieces of equipment (e.g., augers and long cable).
3. 55-gallon drums—These are used to contain the decontamination water upon completion.
4. Metal/plastic buckets/containers—These are for storage and disposal of decontamination solutions (e.g., acids, bases, solvents).

Health and Safety Equipment

I am not a health and safety specialist, so this is not an exhaustive list of what you should bring with you on your jobs. You must read your health and safety plan (HASP) and bring the appropriate equipment for the job you are doing. At a minimum, I keep in my car the following:

1. Safety glasses. I keep sun glasses and clear glasses with me.
2. Hard hat.
3. Steel-Toe boots.
4. Nitrile gloves.
5. First-Aid kit
6. Ear plugs

As I suggested in the beginning of this chapter, put the equipment in plastic totes so that when you go away for the weekend, you can easily remove the sampling gear and fit some luggage into your trunk.

9

Project Management

The information put forth in the following chapter should assist you in managing your projects. This is information that you will not find at a typical management class, but that will help you in the big picture of what you are doing at a site. This chapter is not about creating budgets and managing your time; it is about gaining knowledge to effectively communicate your purpose and goals to your boss, client, and co-workers. This chapter will help you be a well-rounded sampler who can transition more easily from the field to the office.

Risk Based Corrective Action

During the mid 90s, I remember hearing people in the office talk about a girl named Rebecca and I wondered who she was. It turned out that Rebecca was actually RBCA, or Risk Based Corrective Action (ASTM D 2081). Prior to RBCA, many regulatory agencies had no uniform way of communicating to what extent a site had to be cleaned. They might have said that benzene should be remediated to 10 mg/kg in soil, but there was no way to justify it. Industry was becoming frustrated because on one site they could remediate to 10 mg/kg yet on another they would have to remediate to 2 mg/kg. There was little rhyme or reason or consistency, and they could not effectively budget remediation projects because they had no idea what they would have to do. Another problem was that sometimes the same clean-up criteria were being applied to all sites, regardless of use. So a factory site might have to be remediated to the same extent as a park, even though the risks were different.

The people who developed RBCA were toxicologists, engineers, geologists, biologists, and other scientists. Their task was to assess how contaminants affect "human health and the environment." If a contaminant adversely affects human health and the environment, then the contaminant should not be in a place where humans and the environment can be exposed to that contaminant. Further, they assessed the actual risk of various contaminants, and to what quantity one would have to be exposed before it caused harmful affects. Considering the habitats of humans and the environment, they established the concentrations to which they could be exposed in their habitats without "likely" causing harm. They actually use, in many cases, a risk factor of one in a million chances of being harmed.

Consider now our habitats. We live in residential areas, as determined by the zoning laws of our municipalities. We normally do not live in industrial areas because they are zoned commercial/industrial. A factory cannot be set up in a residential area because of zoning laws. Granted, there are cusps where residential zoning might abut an industrial complex. Notwithstanding those circumstances, RBCA figures that if you live in a certain area, you are going to have kids, a garden, and enjoy the outdoors. RBCA also figures you are going to live there for 70 years. Your kids will play in the playgrounds, at school, and in the dirt—they will be exposed to the environment. You will grow vegetables in the garden and eat them, you will wash the vegetables with water, drink the water, breath the air, etc. So, if there are high concentrations of a contaminant in this scenario, chances are you or your kids will be exposed, daily, for 70 years. This is called a residential scenario (the name might vary by state), and RBCA establishes the concentrations of contaminants that could affect you living in this setting. Or, it establishes the contaminant concentrations that will not affect you living in this setting.

The exposure duration actually varies depending on the contaminant. If the contaminant is a carcinogen, then RBCA assumes you will be exposed for 70 years, but only 350 days a year, considering vacations. Other situations might have exposure as low as 6 years. It also depends upon the type of exposure. You can be exposed through dermal contact (on the skin), ingestion (eating), and inhalation (breathing).

If you work in a commercial/industrial area, it is not as likely that kids will be frolicking in the mud for days on end. It is also not likely that you will plant a garden at work, and you will not live there. This scenario assumes you will be in this area 8 hours a day, five days a week, and 250 days a year for up to 25 years. RBCA establishes contaminant concentrations that will not affect you if you work in this setting.

Finally, there are construction worker scenarios. This assumes a construction worker will be excavating contaminated soil and will be there 8 hours a day for 30 days a year. Some scenarios suggest it is a one-year job, others a 40 day job.

Based on these data and myriad physical and chemical properties, RBCA has developed equations that determine the "allowable" exposure concentration of contaminated soil or water. Using these equations, many states have developed "look up" tables that allow you to check the permitted concentration of a chemical in a given scenario. For example, if you analyze a soil sample and find it has benzene at 0.6 mg/kg, then in Illinois, you could live, work, and construct in this area. This is called Tier 1 in RBCA. Looking up the permissible concentrations for your scenario is the first step in evaluating conditions at your site. In this case, there is no exposure so there is no risk. If there is no risk, there is no basis for corrective action. Or, if site concentrations are less than Tier 1, then there is no further remediation required to protect human health and the environment.

Tier 2 gets a little more complex. Assume that soil at a site had a pyrene concentration of 90,000 mg/kg. In Illinois, you could not live nor work there (remediation objective is 60,000 mg/kg). If it is an existing facility that has a concrete floor and paved parking lot, there is really no exposure. It is not likely that the workers are going to start tunneling through the concrete to get to the contaminated soil. If they are standing on the asphalt parking lot they are not exposed to the underlying soil, so there is no exposure to that contaminant. For the company to leave that concentration in their soil, they would have to agree to keep and maintain the concrete floor and asphalt parking lot. Also, any future buyers of the property would likewise have to agree to that restriction. This information would be included with the property deed, and it would become a "deed restriction." That is a Tier 2 analysis. You consider the actual exposure (risk), and if there is none you do not have to remediate, but the corrective action would be maintaining the concrete floor and asphalt parking lot. If there is exposure (e.g., a gravel parking area), then you might have to pave the lot, which could be cheaper than remediation.

Put another way, Tier 2 takes away an aspect of risk. Most people are familiar with the "Fire Triangle," which has oxygen, fuel, and ignition on its sides. If one of the three is removed, then there's no chance for a fire. For risk, you could consider a triangle with source, receptor, and exposure pathway on its sides. Source is the contamination at a site. The receptor is the human or the aspect of the environment that could be exposed to the contamination. The exposure pathway is the route that the contamination could take to affect the receptor. For example, gasoline is spilled and is now a source. Someone (the receptor) is standing near the puddle of gasoline, which is evaporating. The receptor can smell (inhalation) the gasoline fumes; therefore, inhalation is the exposure pathway. You have all three sides of the triangle, so there is risk. If you take away one side of the triangle (clean up the gasoline, evacuate people from the area, or have the receptors don appropriate respirators), then there's no risk.

However, there are other exposure pathways (routes to humans). If the water at a site is contaminated, there are other options you can implement. For example, many municipalities have passed ordinances prohibiting use of groundwater, thereby preventing exposure to contaminated water by drinking. If contaminant inhalation is an issue, you can construct a relatively impermeable barrier between the contaminant and the people breathing. I have been to a building that had an exhaust system beneath the first floor that carried fumes from the ground to the roof. Here, the potential exposure is inhalation; the corrective action to prevent risk is the exhaust system. These are all Tier 2 corrective action scenarios.

A risk assessment uses numerous equations to evaluate what risk is present. Besides affecting site conditions, Tier 2 allows you to use select equations to assess risk. There are equation variables that account for various aspects of the site, including soil characteristics (e.g., actual soil pH, carbon content, porosity, grain size, hydraulic conductivity, soil particle density, hydraulic gradient, permeability), weather, source size, and contaminant characteristics. By providing specific information in these equations, as opposed to default values, you might be able to demonstrate that risk is diminished or eliminated at a site. For example, if the default value for carbon in soil is 6,000mg/kg and the actual site condition is 15,000mg/kg, then the soil at the site would adsorb more organic contaminants than allowed for with the default value, thereby decreasing the risk.

Tier 3 analyses are more complex. They are a full risk assessment in which all equations are open for adjustment, including those under Tier 2.

In addition, you can implement sophisticated groundwater flow models of fate (how a chemical changes in time) and transport (how a chemical moves) specific to the site. You can use toxicological data from sources other than the regulatory defaults provided in many remediation programs. You can consider the land use if it is substantially different from the assumptions of residential or industrial/commercial. Primarily, however, the risk factor is adjusted from 1 in 1,000,000 to 1 in 10,000. Of course, these steps would be taken after regulatory review and approval.

That summarizes the various approaches to remediation using RBCA. The ASTM standard is a guide, and many regulatory agencies use it, or variations of it, in the remediation programs. Please be advised that RBCA has many "sister" guides in ASTM that complement her. There are ASTM guides for fate and transport models, site characterization, groundwater models, RBCA for petroleum release sites, and numerous others. They are valuable resources. You can also consider soil screening levels and other risk calculations provided by the EPA. These can be found at www.epa.gov/superfund. Do not be daunted by the apparent detail and complexity of these standards (or any ASTM standard). You

Career Tip—ASTM standards, guides, and practices are written by various committees comprised of professionals in that field. A great way to understand these guides is to get on the committee and write them. Go to *www.ASTM.org* for information about membership and joining committees.

may not understand them fully at first, but after reading and using them several times they will become more easy to understand.

Oversight Relations

In this business, especially when working on government projects (e.g., Superfund), there may be an oversight contractor hired by the agency to assure that you are doing your job properly. In my experience, I have never had a problem with the oversight contractor. The key to a good relationship is understanding that you both have a job to do and to respect what each is doing. Neither is out to get the other, and neither is competing against the other. When I first start a job with an oversight contractor, I always take the time to explain what our plan is. This includes what we are sampling first, what our daily schedule is (time start and stop), how long the project is expected to last, where we are staying (if an out-of-town job), etc. I also state that if they see anything I am doing as questionable, to let me know immediately. If appropriate, I would adjust my method in the field. The worst thing to have happen is to complete a two week sampling job only to have the oversight submit a report to the agency stating that none of the samples are valid due to improper sampling methods. We would then be re-sampling. It is better to get the questions out in the open as soon as possible.

I worked as oversight on a landfill construction project where I was doing QA/QC. I learned so much from the people whom I was watching. On a Superfund project where I was watched, a consultant provided oversight over several years as sampling events continued. Years later he told me that he learned a great deal from our company. My point is that the person who is watching does not necessarily know more than you, so there is no need to be nervous. They are making sure that the samples collected are representative of the conditions sampled, and want to ensure that you are not filling the sample containers with distilled water or some such thing.

Split samples

On some sites it is necessary to split samples with the oversight contractor. This is another method of validating data. The oversight returns the samples they collect to their lab and you send your samples to your lab. Then the sample results are compared to assess similarity. For split samples, find out ahead of time if that the contractor will collect splits and determine at what locations they will be collected. It is not appropriate for the contractor to pull out a cooler of samples just as you start filling yours. It is even less appropriate for him to pull out the cooler after you have already collected yours. However it happens, be sure to document it in your field book.

When collecting split samples, you alternate between one another's containers. Fill the VOCs of yours, then his. Then fill the SVOCs of yours, then his. Continue this alternating until all parameters are collected. This will provide better data than collecting all of your samples first, then his. You will be comparing his VOC results to yours, so you want them and all other parameters taken as closely together as possible. If you have completed your samples and the oversight says he now wants splits, you should contact respective project managers to discuss the implications of this scenario.

Who's the Boss?

The environmental business is relatively small. It seems that everyone knows someone at some other firm. This is because there is a lot of movement between firms and related businesses. It is important that you keep this in mind when working with others because you never know how your relationships will change throughout your career. I worked for a consultant for several years and made some good friendships. When the office was closed, we split up and went our separate ways. Good friends became competitors, customers, and clients. The environmental field is a small field. The people you loathe may be across from you during your next interview, so do not burn any bridges.

Project Staffing

On large sampling events, there are usually teams of samplers. A team may vary from two to four people. There are specific jobs that have to be conducted throughout the sample collection. When you are at a groundwater well, someone has to purge the well, and someone has to prepare the sample containers (set up, label, arrange). The team should consist of people who like to do the various jobs. I like to separate them into "neck-up" jobs (jobs that require you to use your head) and "neck-down" jobs (that require you to use your muscles). Typically, the person with the most experience will be the neck-up person and the less experienced will be the neck-down person.

Keep in mind that some people will have preferences as to what they like to do. To this day, I would rather bail a well than label a sample container and I'd rather decontaminate equipment than fill out a Chain of Custody form. I am a neck-down sort of guy. Fortunately, many people with whom I have worked have preferred the neck-up work. That makes for a good team. However, at some point, the neck-down person has to be trained to do the neck-up work. That new person will naturally progress to a team leader and will be responsible to perform the neck up work. During a sampling event, the roles should reverse so as to train those who are new, which puts a lot of responsibility on the trainer.

He will have to do the neck-down work, then check that all the sample containers are labeled properly, the COC is filled out properly, etc. .

Hazardous What?

There are several misused terms in the environmental industry. Many people interchange the terms hazardous waste, hazardous substance, hazardous material, hazardous pollutant, and hazardous chemical, yet each has its own definition based upon federal regulations. Likewise, the term toxic is batted about with reckless abandon. Toxic waste is always an issue in the environment, but from a regulatory standpoint, it does not exist. There are wastes that exhibit toxic characteristics, however there is no toxic waste. In a sense, I am getting caught up in semantics, perhaps from the general public's point of view, but within our industry it makes a big difference what you call the "stuff" with which we work.

The regulatory agencies or laws define this hazardous and toxic material. The EPA has authority over most of these regulations however; the Department of Transportation and OSHA are involved, as well. Without getting into the minutiae of the laws, let me give an explanation of select laws and how they (see Table 9.1 for a summary of this discussion) affect a chemical's definition.

When a chemical is manufactured or imported, it must be federally-approved through the Toxic Substances Control Act (TSCA). TSCA maintains an inventory of approved chemical substances. The chemicals on this list are, therefore, chemical substances. TSCA also limits the use of several substances including PCBs, asbestos, dioxins, lead-based paint, nitrosating agents, hexavalent chromium, radon, and chlorofluorocarbons. In 1980, the Comprehensive Environmental Response, Compensation, and Liability Act (CERCLA) was developed (also known as Superfund), and it includes a list of hazardous substances (see 40 CFR Chapter I Part 302.4). There are literally hundreds of hazardous substances, and they include the hazardous wastes from the Resource Conservation and Recovery Act (RCRA). Hazardous wastes are defined in RCRA as a solid waste (which includes liquids) that exhibits one or more of the following: ignitability, toxicity, corrosivity, or reactivity (see 40 CFR 261.3 for a more detailed and accurate definition of hazardous waste).

If you store certain quantities of hazardous substances (Table 302.4) or extremely hazardous substances (EHS) (40 CFR Part 355, Appendices A and B), then you must report them to various authorities, as prescribed by the Superfund Amendments and Re-authorization Act (SARA), specifically, the Emergency Planning and Community Right-to-Know Act (EPCRA). If a certain quantity of a hazardous substance or extremely hazardous substance is released, then you are re-

Table 9.1. Summary of Environmental Laws

Term	Use	Applicable Law		Notes
Hazardous Substance	Release	CERCLA	40 CFR - Chapter I - Part 302	These drive site remediaton.
Hazardous Substance	Storage	EPCRA	Sections 311-312	You have to let people know what you're storing on your property. This is a result of the Bhopal, India, release in 1984.
Extremely Hazardous Substance	Storage	EPCRA	40 CFR Part 355, Appendices A and B	
Hazardous Chemical	Health and Safety Exposure	OSHA	29CFR 1910.1000	This is designed to protect you while you work.
Hazardous Material	Transportation	DOT	49 CFR 172.101	These apply when you transport the material, and they require manifests.
Toxic or Chemical Substance	Manufacture or Import	TSCA	TSCA	This prevents people from manufacturing or importing banned substances.
Toxic or Priority Pollutant	Discharge to navigable water	CWA	40 CFR Part 129.4 or 40 CFR 122 and 132.	
Hazardous Air Pollutant	Release to Air	CAA	Section 112.	188 pollutants.
Criteria Pollutant	Release to Air	CAA	Section 108	Six criteria pollutants are NOx, SOx, lead, CO, ozone, particulates.
Hazardous Waste	Disposal	RCRA	40 CFR 261	Needs to exhibit ignitability, reactivity, toxicity, or corrosivity.

quired to report it to certain regulatory bodies. The need to report the release of a hazardous substance or extremely hazardous substance is dependent upon the reportable quantity (RQ), which is provided in the lists of these substances.

Now if select hazardous or extremely hazardous substances are released to the air, then they become hazardous air pollutants (HAPs), as regulated by the Section 112 of the Clean Air Act (CAA). There are currently 188 HAPs (see *http://www.epa.gov/ttn/atw/188polls.html*) on the list. In addition to the 188 HAPs, there are six criteria pollutants (NOx, SOx, lead, particulate, CO, and ozone).

If the EHS or hazardous substances are discharged into a water body, then they are either toxic pollutants (40 CFR Part 129.4) or pollutants (40 CFR 122 and 132), according to the Clean Water Act. It is interesting in that if you have a National Pollutant Discharge Elimination System (NPDES) permit, then you are releasing pollutants into a water body. However, if you have released a pollutant and have to monitor the water body for the discharged pollutants, then you have to look for hazardous substances again (40 CFR Part 116). If someone is going to drink the water, then they have to make sure the water meets the Maximum Contaminant Levels (MCL) for drinking water (40 CFR Part 141 Appendix A).

If you are working around chemicals and might be exposed to them, then you have to concern yourself with hazardous chemicals according to the Occupational Safety and Health Act (OSHA) (29 CFR 1910.1200). Permissible Exposure Limits for individual hazardous substances have been developed by OSHA and can be found at 29 CFR 1910.1000. It is OSHA that requires health and safety labels on chemicals (e.g. the blue, yellow, red, and white diamonds).

If you are going on the road, in the air, or on water with the hazardous substances, they have just become hazardous materials according to the Department of Transportation (DOT). They regulate the transport of these materials. They are the authority, through the Hazardous Materials Transportation Act (HMTA), that requires labels and manifests when materials go on the road. Their labels include flammable, corrosive, and poisonous. Their regulations can be found at *http://.hazmat.dot.gov/rules.htm* (49CFR 172). The list of hazardous materials can be found at 49CFR Part 172 Subpart B, 172.101.

Take the potential lifespan of benzene as an example. Benzene is a constituent of gasoline and it is produced as a virgin product for other industrial purposes. In the manufacturing process, benzene is a chemical substance regulated under TSCA. However, since they store it on site, they have to report to authorities that they are storing benzene. This is reported under EPCRA, so it is a hazardous substance. When they deliver benzene to the end user, it is sent via tanker truck. Because it is being transported, that EPCRA hazardous substance has become a hazardous material, which is regulated by the Department of Transportation. When it arrives at the next manufacturer, it has become a hazardous substance under EPCRA again. Let's say this manufacturer spills a drum of benzene. There has been a release, so the benzene is a hazardous substance under CERCLA. The benzene is flowing down the floor drain into a river. They have a NPDES permit, so the discharged substance is now a pollutant. Now the coast guard has to be called because they have jurisdiction on the water.

Now imagine that some of the benzene has flowed into the nearby gravel parking lot so you get a contractor to clean up the benzene. They have to wear personal protective equipment to prevent exposure to a hazardous chemical, as regulated under OSHA. They excavate the soil that is contaminated with the hazardous substance (CERCLA release) and put it into a roll-off box. It has to be disposed of, so it is now a hazardous waste (likely ignitable and toxic), as regulated under RCRA. However, you still have to get it to the landfill so it is being driven on the highway as a hazardous material, again. Because benzene is so volatile, you decide to let it volatilize rather than clean up the soil, which is not an allowable practice, but this is a hypothetical. Since it is going into the air, it has become a hazardous air pollutant (HAP), as regulated under the CAA.

To the lay person watching the evening news, it does not matter what they call it (often toxic waste), just as long as they clean it up. For those of us in the field, it is very important that we know what to call the "stuff" with which we work. It takes time and study, but can cause much confusion if we do not call it by its proper name.

Client Needs

Never promise more than you can deliver, and always give the clients what they need, not necessarily what they want. If they do not know what they need, it is up to you to educate them by telling them why they need what you are going to give them and how it will benefit them.

The Written Report

If you are an environmental consultant, it is important that you recognize the end product of your sampling efforts: the written report. After all the wells are installed, developed, and sampled, and after all the equipment is decontaminated, the report of analytical results and what they mean must be written. The object for which the client ultimately pays is a report. It will typically have text, tables, and figures, which are created on a word processor program, spreadsheet program, and CAD program, respectively. Therefore, to be effective in the office, you should learn to use these computer tools effectively. If you can master them, that is even better. You can learn these tools at the local community college, park district, library, or even management training seminar. If you do not learn to use these tools, there will come a time when you will have to stay late to complete a report and no one else in the office will be around, so you will be forced to learn the hard way. The better your report appears, the easier it is for the client and regulators to read it.

Remember, the client may have paid $50,000 for the analytical, time in the field, drill rig, etc., but all she has to show for it is the two-inch report you and your colleagues put together. You may have done everything on time and under budget, but if the final product (report) is a mess, then the client will be unimpressed. She has to bring the report to her bosses, the regulators, lawyers, etc. If they cannot make sense of it, then you have not done your job. It not only has to look good (good formatting, neat clear tables, and useful figures), it has to read well and be technically correct. The next step, therefore, is to learn to write.

Learning to write well can be frustrating and embarrassing. The best way to learn is to have someone who can write better than you review what you have written. This can be done in the office with your letters and reports, or you can take writing classes at a local college or at seminars.

Good writing includes spelling and grammar. (But keep in mind that "spell check" will not catch everything. For instance, the following sentence passes through spell check untouched: Their are for well's on they're sight and there sampling too tomorrow.) When I have a question about something I have writtten, I call my friend who is an English professor, or look it up in a grammar book. Remember that while you may not care that much about writing, clarity and precision are important. If your clients see a lot of mistakes in your writing, they may begin to question the overall accuracy of your work.

Your client and the regulators have to read the report, so it has to make sense to them. It can frustrate them to no end when there are spelling and grammar mistakes, and they may begin to question your professionalism both on paper and in the field. However, while spelling and grammar are important, if the text does not flow or is confusing, then you have still not done your job. You have to logically lead the reader from the reasons for the sampling event, through the event, to the results, to the conclusions, then to the recommendations.

Tables and Figures

Tables and figures have to clearly convey information from the written report. They are intended to clarify and summarize information, not confuse the reader. Regarding tables, the client should be able look at them and understand that which is being presented. If there are symbols used on the table, provide a footnote to them. You may have used these symbols for years, but someone down the line may not know what they mean. Your client might, but her boss might not. Your job is to make it easy for everyone to understand, not to make it easy on yourself.

I am an advocate of color tables. There are numerous data to present on a table, and using colors can easily demonstrate analytical results. For example, if an analyte is less than the Tier 1 residential remediation objective (RO) then it is represented in green. If it is greater than residential but less than commercial/industrial, then it is blue. If it is greater than commercial/industrial, then it is represented in yellow. I do not recommend using red for the elevated concentrations, as it may give people an unnecessary jolt.

The goal of a table is not to get as much information on one sheet as possible. More is not better and more information that is not needed is worse. Only put on the primary tables key information that the client needs to make decisions. For example, if you sample 20 wells and 14 of them were non-detect, do not intersperse the contaminated wells within the rows and rows of non-detects. Instead, make a separate table with the six wells.

Keep in mind that your client might not be able to see small figures well, so create tables that are easy for everyone to read. Give the clients what they need and are able to read.

There are some basic rules regarding figures. First, have a north arrow that points either up or to the left. Never put south on the top of your figure and never put north to the right. If you have a landscape drawing in a bound report, the reader will tend to put the left side to the top and the right side to the bottom. This holds true in the publishing industry, too (magazines, books, etc.). Maintain this rule in your report and drawings. Also, be sure to bind the top of the landscape drawing and not the bottom.

While this isn't a rule, a good suggestion is to use a normal scale, not some bizarre one inch to 43 feet measurement. Try to use one inch to 10 or 50, as those are easier to handle than 20, 30, 40, and 60. Also, include a bar scale so if the drawing is photocopied, you still have a true representation rather than a number. Photocopies skew drawings, thereby affecting scales. Metric scales are very easy to use, as they are actually ratios. For example, a scale might be 1:1,000. That means one centimeter on the drawing represents 1,000 centimeters at the site. So, one centimeter on the figure is ten meters at the site. Likewise, one inch on the figure is 1,000 inches at the site, but then you have to convert that to feet.

The client should be able to pick up a drawing and identify from where samples were collected. It is a bonus if the drawing also shows select analytical results that will assist the reader in understanding the extent of contamination. Color can also assist with displaying contamination. Colors that are linked to the above-referenced drawings will assist beautifully in this demonstration. A simple black and white drawing with some cross-hatching is blasé. Get some color, get some dimension, and use some creativity.

Index